Ordinary Differential Equations

The MIT Press   Cambridge, Massachusetts, and London, England

Ordinary Differential Equations

V. I. Arnold

Translated and Edited by Richard A. Silverman

Fifth printing, 1987

Printed and bound in the United States of America by Halliday Lithograph.

Library of Congress Cataloging in Publication Data

Arnold, Vladimir Igorevich.
  Ordinary differential equations.

  Translation of Obyknovennye differentsial'nye Uravneniya.
  1. Differential equations. I. Title.
QA372.A713   515'.352   73-6846
ISBN 0-262-01037-2 (hardcover)
ISBN 0-262-51018-9 (paperback)

Contents

Preface

In selecting the subject matter of this book, I have attempted to confine myself to the irreducible minimum of absolutely essential material. The course is dominated by two central ideas and their ramifications: The theorem on rectifiability of a vector field (equivalent to the usual theorems on existence, uniqueness, and differentiability of solutions) and the theory of one-parameter groups of linear transformations (i.e., the theory of linear autonomous systems). Accordingly, I have taken the liberty of omitting a number of more specialized topics usually included in books on ordinary differential equations, e.g., elementary methods of integration, equations which are not solvable with respect to the derivative, singular solutions, Sturm-Liouville theory, first-order partial differential equations, etc. The last two topics are best considered in a course on partial differential equations or calculus of variations, while some of the others are more conveniently studied in the guise of exercises.

On the other hand, the applications of ordinary differential equations to mechanics are considered in more than the customary detail. Thus the pendulum equation appears at the very beginning of the book, and the efficacy of various concepts and methods introduced throughout the book are subsequently tested by applying them to this example. In this regard, the law of conservation of energy appears in the section on first integrals, the "method of small parameters" is deduced from the theorem on differentiation with respect to a parameter, and the theory of linear equations with periodic coefficients leads naturally to the study of the swing ("parametric resonance").

Many of the topics dealt with here are treated in a way drastically different from that traditionally encountered. At every point I have tried to emphasize the geometric and qualitative aspect of the phenomena under consideration. In keeping with this policy, the book is full of figures but contains no formulas of any particular complexity. On the other hand, it presents a whole congeries of fundamental concepts (like phase space and phase flows, smooth manifolds and tangent bundles, vector fields and one-parameter groups of diffeomorphisms) which remain in the shadows in the traditional coordinate-based approach. My book might have been considerably abbreviated if these concepts could have been regarded as known, but unfortunately they are not presently included in courses either on analysis or geometry. Hence I have been compelled to present them in some detail, without assuming any background on the part of the reader beyond the scope of the standard elementary courses on analysis and linear algebra.

This book stems from a year's course of lectures given by the author to students of mathematics at Moscow University during the academic years 1968–1969 and 1969–1970. In preparing the lectures for publication I have received great assistance from R. I. Bogdanov. I wish to thank him and all my colleagues and students who have commented on the preliminary mimeograph edition of the book (Moscow University, 1969). I am also grateful to D. V. Anosov and S. G. Krein for their careful reading of the manuscript.

V. I. Arnold

Frequently Used Notation

R the set (group, field) of real numbers.

C the set (group, field) of complex numbers.

Z the set (group, ring) of integers.

$\varnothing$ the empty set

$x \in X \subset Y$ an element x of a subset X of a set Y.

$X \cup Y, X \cap Y$ the union and intersection of the sets X and Y.

$X \backslash Y, X \backslash a$ the set of elements in X but not in Y, the set X minus the element $a \in X$.

$f: X \to Y$ a mapping f of a set X into a set Y.

$x \mapsto y$ the mapping carries the point x into the point y.

$f \circ g$ the product (composition) of two mappings (g is applied first).

$\exists, \forall, \Rightarrow$ there exists, for every, implies.

Theorem 0.0 the unique theorem in Sec. 0.0.

$\blacksquare$ end of proof symbol.

* an optional (more difficult) problem or theorem.

$\mathbf{R}^n$ a linear space of dimension n over the field $\mathbf{R}$.

$\mathbf{R}_1 \dotplus \mathbf{R}_2$ the direct sum of the spaces $\mathbf{R}_1$ and $\mathbf{R}_2$.

$GL(\mathbf{R}^n)$ the group of linear automorphisms of $\mathbf{R}^n$.

One can consider other structures as well in the set $\mathbf{R}^n$, e.g., affine or Euclidean structure, or even the structure of the direct product of n lines. This will usually be spelled out explicitly, by referring to "the affine space $\mathbf{R}^n$," "the Euclidean space $\mathbf{R}^n$," "the coordinate space $\mathbf{R}^n$," and so on.

Elements of a linear space are called *vectors*, and are usually denoted by boldface letters ($\mathbf{v}, \boldsymbol{\xi}$, etc.). Vectors of the space $\mathbf{R}^n$ are identified with sets of n numbers. For example, we write $\mathbf{v} = (v_1, \ldots, v_n) = v_1\mathbf{e}_1 + \cdots + v_n\mathbf{e}_n$, where the set of n vectors $\mathbf{e}_1, \ldots, \mathbf{e}_n$ is called a *basis* in $\mathbf{R}^n$. The norm (length) of the vector $\mathbf{v}$ in the Euclidean space $\mathbf{R}^n$ is denoted by $|\mathbf{v}|$ and the scalar product of two vectors $\mathbf{v} = (v_1, \ldots, v_n), \mathbf{w} = (w_1, \ldots, w_n) \in \mathbf{R}^n$ by $(\mathbf{v}, \mathbf{w})$. Thus

$$(\mathbf{v}, \mathbf{w}) = v_1 w_1 + \cdots + v_n w_n,$$
$$|\mathbf{v}| = \sqrt{(\mathbf{v}, \mathbf{v})} = \sqrt{v_1^2 + \cdots + v_n^2}.$$

We often deal with functions of a real parameter t called the *time*. Differentiation with respect to t (giving rise to a *velocity* or *rate of change*) is usually denoted by an overdot, as in $\dot{x} = dx/dt$.

1 Basic Concepts

1. Phase Spaces and Phase Flows

The theory of ordinary differential equations is one of the basic tools of mathematical science. The theory allows us to study all kinds of evolutionary processes with the properties of *determinacy*, *finite-dimensionality*, and *differentiability*. Before undertaking exact mathematical definitions, we consider a few examples.

1.1. Examples of evolutionary processes. A process is said to be *deterministic* if its entire future course and its entire past are uniquely determined by its state at the present instant of time. The set of all possible states of a process is called its *phase space*.

Thus, for example, classical mechanics considers the motion of systems whose past and future are uniquely determined by the initial positions and initial velocities of all points of the system. The phase space of a mechanical system is just the set whose typical element is a set of instantaneous positions and velocities of all particles of the system.

The motion of particles in quantum mechanics is not described by a deterministic process. Heat propagation is a semi-deterministic process, in that its future is determined by its present but not its past.

A process is said to be *finite-dimensional* if its phase space is finite-dimensional, i.e., if the number of parameters required to describe its state is finite. Thus, for example, the classical (Newtonian) motion of a system consisting of a finite number of particles or rigid bodies comes under this heading. In fact, the dimension of the phase space of a system of n particles is just $6n$, while that of a system of n rigid bodies is just $12n$. As examples of processes which cannot be described by using a finite-dimensional phase space, we cite the motion of fluids (studied in hydrodynamics), oscillations of strings and membranes, and the propagation of waves in optics and acoustics.

A process is said to be *differentiable* if its phase space has the structure of a differentiable manifold and if its change of state with time is described by differentiable functions. For example, the coordinates and velocities of the particles of a mechanical system vary in time in a differentiable manner, while the motions studied in shock theory do not have the differentiability property. By the same token, the motion of a system in classical mechanics can be described by using ordinary differential equations, while other tools are used in quantum mechanics, the theory of heat conduction, hydrodynamics, the theory of elasticity, optics, acoustics, and the theory of shock waves.

The process of radioactive decay and the process of reproduction of bac-

teria in the presence of a sufficient amount of nutrient medium afford two
more examples of deterministic finite-dimensional differentiable processes.
In both cases the phase space is one-dimensional, i.e., the state of the process
is determined by the quantity of matter or the number of bacteria, and in
both cases the process is described by an ordinary differential equation.

It should be noted that the form of the differential equation of the process
and the very fact that we are dealing with a deterministic finite-dimensional
differentiable process in the first place, can only be established experimentally—and hence only with a certain degree of accuracy. However, this
state of affairs will not be emphasized at every turn in what follows; instead,
we will talk about real processes as if they actually coincided with our
idealized mathematical models.

1.2. Phase flows. An exact formulation of the general principles just
presented requires the rather abstract notions of *phase space* and *phase flow*.
To familiarize ourselves with these concepts, we consider an example due
to N. N. Konstantinov where the simple act of introducing a phase space
allows us to solve a difficult problem.

Problem 1. Two nonintersecting roads lead from City A to City B (Fig. 1).
Suppose it is known that two cars connected by a rope of length less than $2l$
manage to go from A to B along different roads without breaking the rope.
Can two circular wagons of radius l whose centers move along the roads in
opposite directions pass each other without colliding?

Solution. Consider the square

$$M = \{(x_1, x_2) : 0 \leqslant x_1 \leqslant 1, 0 \leqslant x_2 \leqslant 1\}$$

Fig. 1   Initial position of the wagons.

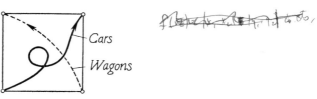

Fig. 2   Phase space of a pair of vehicles.

d = motion determined by (angle of road).

$$\frac{d(A,P)}{d(A,B)} = x_1$$

1st road

2nd road

(Fig. 2). The position of two vehicles (one on the first road, the other on the second road) can be characterized by a point of the square M; we need only let x_i denote the fraction of the distance from A to B along the ith road which lies between A and the vehicle on the given road. Clearly there is a point of the square M corresponding to every possible state of the pair of vehicles. The square M is called the *phase space*, and its points are called *phase points*.

Thus every phase point corresponds to a definite position of the pair of vehicles (apart from their being connected), and every motion of the vehicles is represented by a motion of the phase point in the phase space. For example, the initial position of the cars (in City A) corresponds to the lower left-hand corner of the square ($x_1 = x_2 = 0$), and the motion of the cars from A to B is represented by a curve going to the opposite (upper right-hand) corner of the square. In just the same way, the initial position of the wagons corresponds to the lower right-hand corner of the square ($x_1 = 1$, $x_2 = 0$), and the motion of the wagons is represented by a curve leading to the opposite (upper left-hand) corner of the square. But every pair of curves in the square joining different pairs of opposite corners must intersect. Therefore, no matter how the wagons move, there comes a time when the pair of wagons occupies a position occupied at some time by the pair of cars. At this time the distance between the centers of the wagons will be less than $2l$, and they will not manage to pass each other.

Although differential equations play no role in the above example, the considerations which are involved closely resemble those which will concern us subsequently. Description of the states of a process as points of a suitable phase space often turns out to be extraordinarily useful.

We now return to the concepts of determinacy, finite-dimensionality, and differentiability of a process. The mathematical model of a deterministic process is a *phase flow*, which can be described as follows in intuitive terms: Let M be the phase space and $x \in M$ an initial state of a process, and let $g^t x$ denote the state of the process at time t, given that its initial state is x. For every real t this defines a mapping

$$g^t : M \to M$$

of the phase space into itself. The mapping g^t, called the *t-advance mapping*, maps every state $x \in M$ into a new state $g^t x \in M$. For example, g^0 is the identity mapping which leaves every point of M in its original position. Moreover

$$g^{t+s} = g^t g^s,$$

since the state $y = g^s x$ (Fig. 3), into which x goes after time s, goes after time

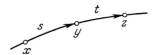

Fig. 3   Change of state of a process in the course of time.

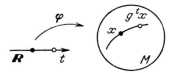

Fig. 4   Motion of a phase point in the phase space M.

t into the same state $z = g^t y$ as the state $z = g^{t+s} x$ into which x goes after time $t + s$.

Suppose we fix a phase point $x \in M$, i.e., an initial state of the process. In the course of time the state of the process will change, and the point x will describe a *phase curve* $\{g^t x, t \in \mathbf{R}\}$ in the phase space M. It is just the family of t-advance mappings $g^t : M \to M$ that constitutes a *phase flow*, with each phase point moving along its own phase curve.

We now turn to precise mathematical definitions. In each case M is an arbitrary set.

Definition. A family $\{g^t\}$ of mappings of a set M into itself, labelled by the set of all real numbers $(t \in \mathbf{R})$, is called a *one-parameter group of transformations* of M if

$$g^{t+s} = g^t g^s \tag{1}$$

for all s, $t \in \mathbf{R}$ and g^0 is the identity mapping (which leaves every point fixed).

Problem 2. Prove that a one-parameter group of transformations is a commutative group and that every mapping $g^t : M \to M$ is one-to-one. ✓

Definition. A pair $(M, \{g^t\})$ consisting of a set M and a one-parameter group $\{g^t\}$ of transformations of M into itself is called a *phase flow*. The set M is called the *phase space* of the flow, and its elements are called *phase points*.

Definition. Let $x \in M$ be any phase point, and consider the mapping

$$\varphi : \mathbf{R} \to M, \qquad \varphi(t) = g^t x \tag{2}$$

of the real line into phase space (Fig. 4). Then the mapping (2) is called the *motion* of the point x under the action of the flow $(M, \{g^t\})$.

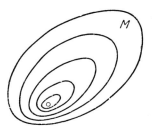

Fig. 5   Phase curves.

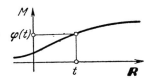

Fig. 6   An integral curve in extended phase space.

Definition. The image of **R** under the mapping (2) is called a *phase curve* of the flow $(M, \{g^t\})$. Thus a phase curve is a subset of phase space (Fig. 5).

Problem 3. Prove that there is one and only one phase curve passing through every point of phase space.

Definition. By an *equilibrium position* or *fixed point* $x \in M$ of a flow $(M, \{g^t\})$ is meant a phase point which is itself a phase curve:

$$g^t x = x \quad \forall t \in \mathbf{R}.$$

The concepts of *extended phase space* and *integral curve* are associated with the graph of the mapping φ. First we recall that the *direct product* $A \times B$ of two given sets A and B is defined as the set of all ordered pairs (a, b), $a \in A$, $b \in B$, while the *graph* of a mapping $f: A \to B$ is defined as the subset of the direct product $A \times B$ consisting of all points $(a, f(a))$, $a \in A$.

Definition. By the *extended phase space* of a flow $(M, \{g^t\})$ is meant the direct product $\mathbf{R} \times M$ of the real t-axis and the phase space M. The graph of the motion (2) is called an *integral curve* (Fig. 6) of the flow $(M, \{g^t\})$.

Problem 4. Prove that there is one and only one integral curve passing through every point of extended phase space.

Problem 5. Prove that the horizontal line $\mathbf{R} \times x$, $x \in M$ is an integral curve if and only if x is an equilibrium position.

Problem 6. Prove that a shift

$$h^s: (\mathbf{R} \times M) \to (\mathbf{R} \times M), \qquad h^s(t, x) = (t + s, x)$$

of extended phase space along the time axis carries integral curves into integral curves.

1.3. Diffeomorphisms. The above definitions formalize the concept of a deterministic process. The corresponding formalization of the concepts of finite-dimensionality and differentiability consists in requiring that the phase space be a *finite-dimensional differentiable manifold* and that the phase flow be a one-parameter group of diffeomorphisms of this manifold.

We now clarify these terms. Examples of differentiable manifolds are afforded by Euclidean spaces and their open sets, circles, spheres, tori, etc. A general definition will be given in Chap. 5, but for the time being it can be assumed that we are talking about an (open) domain of Euclidean space.

By a *differentiable function* $f: U \to \mathbf{R}$ defined in a domain U of n-dimensional Euclidean space $\mathbf{R}^n$ with coordinates $x_1, \ldots, x_n$ we mean an r-fold continuously differentiable function $f(x_1, \ldots, x_n)$ where $1 \leqslant r \leqslant \infty$. In most cases the exact value of r is of no interest and hence will not be indicated; in cases where it is required, we will allude to "r-differentiability" or the function class C^r.

By a *differentiable mapping* $f: U \to V$ of a domain U of n-dimensional Euclidean space $\mathbf{R}^n$ with coordinates $x_1, \ldots, x_n$ into a domain V of m-dimensional Euclidean space $\mathbf{R}^m$ with coordinates $y_1, \ldots, y_m$ we mean a mapping given by differentiable functions $y_i = f_i(x_1, \ldots, x_n)$. This means that if $y_i: V \to \mathbf{R}$ are the coordinates in V, then $y_i \circ f: U \to \mathbf{R}$ are differentiable functions in U $(1 \leqslant i \leqslant m)$.

By a *diffeomorphism* $f: U \to V$ we mean a one-to-one mapping such that both f and $f^{-1}: V \to U$ are differentiable mappings.

Problem 1. Which of the following functions specify a diffeomorphism $f: \mathbf{R} \to \mathbf{R}$ of the line onto the line:
$$f(x) = 2x, \; x^2, \; x^3, \; e^x, \; e^x + x?$$

Problem 2. Prove that if $f: U \to V$ is a diffeomorphism, then the Euclidean spaces with the domains U and V as subsets have the same dimension. *Hint.* Use the implicit function theorem.

Definition. By a *one-parameter group* $\{g^t\}$ *of diffeomorphisms* of a manifold M (which can be thought of as a domain in Euclidean space) is meant a mapping

$$g: \mathbf{R} \times M \to M, \qquad g(t, x) = g^t x, \qquad t \in \mathbf{R}, \qquad x \in M$$

of the direct product $\mathbf{R} \times M$ into M such that
1) g is a differentiable mapping;
2) The mapping $g^t: M \to M$ is a diffeomorphism for every $t \in \mathbf{R}$;
3) The family $\{g^t, t \in \mathbf{R}\}$ is a one-parameter group of transformations of M.

Example 1. $M = \mathbf{R}$, $g^t x = x + vt$ $(v \in \mathbf{R})$.

Remark. Property 2 is a consequence of properties 1) and 3) (why?).

1.4. Vector fields. Let $(M, \{g^t\})$ be a phase flow, given by a one-parameter group of diffeomorphisms of a manifold M in Euclidean space.

Definition. By the *phase velocity* $\mathbf{v}(x)$ of the flow g^t at a point $x \in M$ (Fig. 7) is meant the vector representing the velocity of motion of the phase point, i.e.,

$$\frac{d}{dt}\bigg|_{t=0} g^t x = \mathbf{v}(x). \tag{3}$$

The left-hand side of (3) is often denoted by $\dot{x}$. Note that the derivative is defined, since the motion is a differentiable mapping of a domain in Euclidean space.

Problem 1. Prove that $\dfrac{d}{dt}\, g^t x = \dfrac{d\,[g^t x]}{d(t-\tau)} \cdot \dfrac{d(t-\tau)}{dt} = \dfrac{d\,[g^{t-\tau}g^{\tau}x]}{d(t-\tau)} = \dfrac{d}{dt}\,g^{t}g^{\tau}x.$

$\dfrac{d}{dt}\bigg|_{t=\tau} g^t x = \mathbf{v}(g^\tau x),$ Now evaluate the two ends at $t=0$.

i.e., that at every instant of time the vector representing the velocity of motion of the phase point equals the vector representing the phase velocity at the very point of phase space occupied by the moving point at the given time.

Hint. See (1). The solution is given in Sec. 3.2.

If $x_1, \ldots, x_n$ are the coordinates in our Euclidean space, so that $x_i : M \to \mathbf{R}$, then the velocity vector $\mathbf{v}(x)$ is specified by n functions $v_i : M \to \mathbf{R}$, $i = 1, \ldots, n$, called the *components* of the velocity vector:

$$v_i(x) = \frac{d}{dt}\bigg|_{t=0} x_i(g^t x).$$

Problem 2. Prove that v_i is a function of class C^{r-1} if the one-parameter group $g : \mathbf{R} \times M \to M$ is of class C^r.

Definition. Let M be a domain in Euclidean space with coordinates $x_1, \ldots, x_n$ ($x_i : M \to \mathbf{R}$), and suppose that with every point $x \in M$ there is associated the vector $\mathbf{v}(x)$ emanating from x. Then this defines a *vector field* $\mathbf{v}$ on M, specified in the x_i coordinate system by n differentiable functions $v_i : M \to \mathbf{R}$.

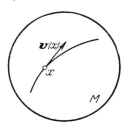

Fig. 7   The phase velocity vector.

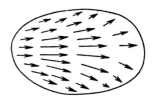

Fig. 8   A vector field.

Thus the aggregate of phase velocity vectors forms a vector field on the phase space M, namely the phase velocity field $\mathbf{v}$ (Fig. 8).

Problem 3. Prove that if x is a fixed point of a phase flow, then $\mathbf{v}(x) = 0$.

x fixed $\Rightarrow$ $g^t x = x$ is constant $\Rightarrow$ $\frac{d}{dt} g^t x = 0$

A point at which a vector of a given vector field vanishes is called a *singular point* of the vector field.† Thus the equilibrium positions of a phase flow are singular points of the phase velocity field. The converse is true, but is not so easy to prove.

1.5. The basic problem of the theory of ordinary differential equations. The basic problem of the theory of ordinary differential equations consists in investigating 1) one-parameter groups $\{g^t\}$ of diffeomorphisms of a manifold M, 2) vector fields on M, and 3) the relations between 1) and 2). We have already seen that the group $\{g^t\}$ defines a vector field on M, i.e., the field of the phase velocity $\mathbf{v}$, in accordance with formula (3). Conversely, it turns out that a vector field $\mathbf{v}$ uniquely determines a phase flow (under certain conditions to be given below).

Speaking informally, we can say that the vector field of the phase velocity gives the *local law of evolution* of a process, and that the task of the theory of ordinary differential equations is to reconstruct the past and predict the future of the process from a knowledge of this local law of evolution.

1.6. Examples of vector fields.

Example 1. It is known from experiment that *the rate of radioactive decay is proportional to the amount x of matter present at any given time.* Here the phase space is the half-line

$$M = \{x : x > 0\}$$

(Fig. 9), and the indicated experimental fact means that

$$\dot{x} = -kx, \qquad \mathbf{v}(x) = -kx, \qquad k > 0, \tag{4}$$

† Note that the components of the field have no singularities at a singular point, and in fact are continuously differentiable. The term "singular point" stems from the fact that the direction of the vectors of the field changes near such a point, in general discontinuously.

$\dfrac{dx}{dt} = -kx$

$\int \dfrac{1}{x} \dfrac{dx}{dt} \, dt = \int -k \, dt$

$\ln x = -kt + c$

$x = C_0 e^{-kt}$

$\phi : \mathbb{R} \to M : t \mapsto x(0) e^{-kt}$

Integral curve in the graph of $x = x_0 e^{-kt}$

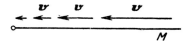

Fig. 9   The phase space of radioactive decay.

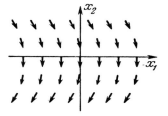

Fig. 10   The phase plane for vertical fall.

i.e., the vector field **v** on the half line is directed toward 0 and the magnitude of the phase velocity vector is proportional to x.

Example 2. It is known from experiment that *the reproduction rate of a colony of bacteria supplied with enough food is proportional to the quantity x of bacteria present at any given time.* Again M is the half-line $x > 0$, but the vector field differs in sign from that of the previous example:

$$\dot{x} = kx, \qquad \mathbf{v}(x) = kx, \qquad k > 0. \tag{5}$$

Note that equation (5) corresponds to growth, with the increase proportional to the number of individuals present.

Example 3. One can imagine a situation where *the increase is proportional to the total number of pairs present*, i.e.,

$$\dot{x} = kx^2, \qquad \mathbf{v}(x) = kx^2 \tag{6}$$

(this situation is more readily encountered in physical chemistry than in biology). Later we will see the catastrophic consequences of the excessively rapid law of growth (6).

Example 4. Vertical fall of a particle to the ground (starting from not too great an initial height) is described experimentally by Galileo's law, which asserts that the acceleration is constant. Here the phase space M is the plane (x_1, x_2), where x_1 is the height and x_2 the velocity, while Galileo's law is expressed by formulas like (3), namely

$$\dot{x}_1 = x_2, \qquad \dot{x}_2 = -g \tag{7}$$

($-g$ is the acceleration due to gravity). The corresponding vector field of the phase velocity has components $v_1 = x_2, v_2 = -g$ (Fig. 10).

[handwritten annotations:] Integral curve for $(x > 0)$ Ex 2 is just ... as for Ex 1. Ex 3: $\frac{dx}{dt} = kx^2 \Rightarrow x = -\frac{1}{kt - c}$, $x > 0$, so $C = \frac{1}{x_0}$ where x_0 is x-int. Integral curves

Example 5. The small oscillations of a plane pendulum are described by a two-dimensional phase plane with coordinates x_1 and x_2, where x_1 is the angle of deviation from the vertical, x_2 is the angular velocity, and M is a neighborhood of the origin of coordinates. According to the laws of mechanics, the acceleration is proportional to the angle of deviation. Thus

$$\dot{x}_1 = x_2, \qquad \dot{x}_2 = -kx_1, \qquad k = l/g, \tag{8}$$

where l is the length of the pendulum and g is the acceleration due to gravity. In other words, the vector field of the phase velocity has components $v_1 = x_2, v_2 = -kx_1$. The origin is a singular point of this vector field (Fig. 11).

Example 6. A more exact description of the (not necessarily small) oscillations of the pendulum leads to the law

$$\dot{x}_1 = x_2, \qquad \dot{x}_2 = -k \sin x_1. \tag{9}$$

The corresponding vector field in the phase plane with coordinates x_1, x_2 is just

$$v_1 = x_2, v_2 = -k \sin x_1$$

(Fig. 12), with singular points $x_1 = m\pi, x_2 = 0$. Note that it is natural to

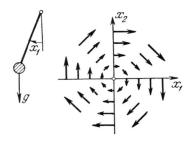

Fig. 11   Small oscillations of a pendulum.

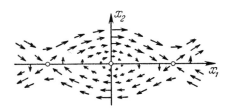

Fig. 12   Phase velocity field of a pendulum.

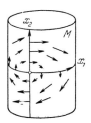

Fig. 13   The cylindrical phase space of a pendulum.

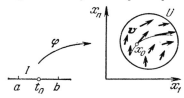

Fig. 14   Solution of the differential equation $\dot{x} = \mathbf{v}(x)$ satisfying the initial condition $\varphi(t_0) = x_0$.

regard the phase space of the pendulum as being the surface of the cylinder $(x_1 \bmod 2\pi, x_2)$ rather than the plane (x_1, x_2), since changing the angle x_1 by 2π does not change the state of the pendulum. The vector field corresponding to (9) can also be regarded as defined on the surface of a cylinder (Fig. 13).

Problem 1. Sketch integral curves for Examples 1–3, and phase curves for Examples 4 and 5.

See bottom of pg 8,9.

2. Vector Fields on the Line

We now show how the operation of integration (with the help of the fundamental theorem of calculus) allows one to solve differential equations determined by vector fields on the line. We begin by introducing some definitions that will be used repeatedly below.

2.1. Solutions of differential equations. Let U be an (open) domain of n-dimensional Euclidean space, and let $\mathbf{v}$ be a vector field in U (Fig. 14). Then by the *differential equation determined by the vector field* $\mathbf{v}$ is meant the equation†

$$\dot{x} = \mathbf{v}(x), \qquad x \in U. \tag{1}$$

† Differential equations are sometimes said to be equations containing unknown functions and their derivatives. This is false. For example, the equation
$$\frac{dx}{dt} = x(x(t))$$
is not a differential equation.

The domain U is called the *phase space* of equation (1).

Definition. By a *solution* of the differential equation (1) is meant a differentiable mapping $\varphi: I \to U$ of the interval $I = \{t \in \mathbf{R}, a < t < b\}$ of the real t-axis $(a = -\infty, b = +\infty$ are allowed) into the phase space such that

$$\frac{d}{dt}\bigg|_{t=\tau} \varphi(t) = \mathbf{v}(\varphi(\tau))$$

for all $\tau \in I$.

In other words, as t varies, the point $\varphi(t)$ must move in U in such a way that its velocity at every instant of time τ equals the vector $\mathbf{v}(x)$ of the field $\mathbf{v}$ at the point $x = \varphi(\tau)$ occupied by the moving point at the given instant. The image of I under the mapping φ is called a *phase curve* of the differential equation (1).

Definition. Suppose the value of a solution $\varphi: I \to U$ of the differential equation (1) at the point t_0, $a < t_0 < b$ equals x_0, i.e., suppose the phase curve goes through the point x_0 at the time t_0. Then φ is said to *satisfy the initial condition*

$$\varphi(t_0) = x_0, \qquad t_0 \in \mathbf{R}, \qquad x_0 \in U. \tag{2}$$

Example 1. If x_0 is a singular point of the vector field, so that $\mathbf{v}(x_0) = 0$, then $\varphi = x_0$ is a solution of equation (1) satisfying the initial condition (2). Such a solution is called an *equilibrium position* or *stationary solution*, and the point x_0 is then also a phase curve.

In general it is impossible to find the solutions of a differential equation explicitly, starting from a knowledge of the vector field. The basic case in which this can be done is the case $n = 1$, i.e., the case of vector fields on the line. We now study this case.

2.2. Integral curves.

Definition. The direct product $\mathbf{R} \times U$ is called the *extended phase space* of equation (1), and the graph of any solution of (1) is called an *integral curve* of (1).

In the case under consideration $(n = 1)$, the extended phase space is a strip $\mathbf{R} \times U$ in the direct product of the t-axis and the x-axis (Fig. 15).

Suppose that through every point (t, x) of extended phase space we draw a straight line whose angle of inclination with the positive t-axis has tangent $\mathbf{v}(x)$. Then the resulting family of straight lines is called the *direction field associated with equation* (1) or simply the *direction field* $\mathbf{v}$.

Every integral curve is tangent to the direction field $\mathbf{v}$ at each of its points. Conversely, every curve tangent at each of its points to the direction $\mathbf{v}$ at the

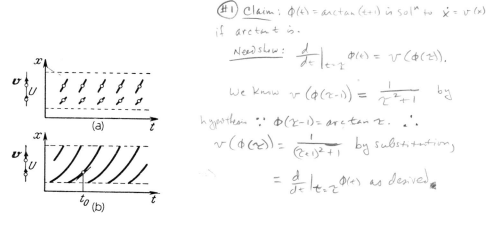

Fig. 15   A direction field (a) and integral curves (b) in extended phase space.

given point is an integral curve (prove this!). *Derivative of fctn induced by graph at x is $v(x)$ which is dy/dt of sol'n.*

A solution of (1) satisfies the initial condition (2) if and only if the corresponding integral curve goes through the point (t_0, x_0). Thus finding the solution of (1) satisfying (2) is equivalent to drawing a curve through (t_0, x_0) which is tangent at each of its points to the direction field **v**.

Note that the slope of the integral curves is the same everywhere along a given horizontal line $x = $ const.

Problem 1. Let $x = \arctan t$ be a solution of equation (1). Prove that $x = \arctan(t + 1)$ is also a solution. *See above*

Hint. The solution is given in Sec. 10.1.

2.3. Theorem. *Let* **v**: $U \to$ **R** *be a differentiable function defined on an interval*

$$U = \{x \in \mathbf{R}: \alpha < x < \beta\}, \qquad -\infty \leqslant \alpha < \beta \leqslant +\infty$$

of the real axis. Then
1) *For every $t_0 \in$ **R**, $x_0 \in U$ there exists a solution φ of equation (1) satisfying the initial condition (2);*
2) *Any two solutions φ_1, φ_2 of equation (1) satisfying (2) coincide in some neighborhood of the point $t = t_0$;*
3) *The solution φ of equation (1) satisfying (2) is such that*

$$
\begin{aligned}
t - t_0 &= \int_{x_0}^{\varphi(t)} \frac{d\xi}{\mathbf{v}(\xi)} \quad &&\text{if } \mathbf{v}(x_0) \neq 0, \\
\varphi(t) &= x_0 \quad &&\text{if } \mathbf{v}(x_0) = 0.
\end{aligned}
\tag{3}
$$

Remark. Since $\mathbf{v}(\xi)$ is a known function, formula (3) allows us to find the function ψ inverse to φ $(t = \psi(x), \varphi(t) = x)$ by quadratures. We can then use the implicit function theorem to find φ. Thus formula (3) leads to the solution of equation (1) subject to the condition (2).

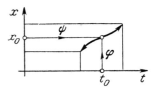

Fig. 16   A solution φ and its inverse function ψ.

2.4. Beginning of the proof of Theorem 2.3.

a) If $\mathbf{v}(x_0) = 0$, let $\varphi(t) \equiv x_0$. Then φ is a solution of (1) and (2) satisfying (3).

b) Let $\mathbf{v}(x_0) \neq 0$, and let φ be a solution of (1) and (2). Then, by the implicit function theorem, the function ψ inverse to φ ($t = \psi(x)$, $\psi(x_0) = t_0$) is defined in a sufficiently small neighborhood of the point x_0 (Fig. 16) and

$$\left.\frac{d\psi}{dx}\right|_{x=\xi} = \frac{1}{\mathbf{v}(\xi)}.$$

Since $\mathbf{v}(x_0) \neq 0$, the function $1/\mathbf{v}(\xi)$ is continuous in a sufficiently small neighborhood of the point $\xi = x_0$, and hence

$$\psi(x) - \psi(x_0) = \int_{x_0}^{x} \frac{d\xi}{\mathbf{v}(\xi)}$$

by the fundamental theorem of calculus. This uniquely defines ψ in a sufficiently small neighborhood of the point $x = x_0$. The function φ inverse to ψ is also uniquely defined in some neighborhood of the point $t = t_0$ by the condition $\varphi(t_0) = x_0$ (the implicit function theorem is applicable since $1/\mathbf{v}(x_0) \neq 0$). Thus any solution of equation (1) subject to the condition (2) satisfies (3) in a sufficiently small neighborhood of the point $t = t_0$, and the uniqueness assertion 2) is proved.

c) We must still verify that the function φ inverse to ψ is a solution of (1) and (2). But

$$\frac{d\varphi}{dt} = \left.\frac{d\psi^{-1}}{dt}\right|_{x=\varphi(t)} = \left.\left(\frac{1}{\mathbf{v}(x)}\right)^{-1}\right|_{x=\varphi(t)} = \mathbf{v}(\varphi(t)), \qquad \varphi(t_0) = x_0,$$

and the theorem is "proved."

Problem 1. Find the gap in the proof.

2.5. Failure of uniqueness. Let $\mathbf{v} = x^{2/3}$, $t_0 = 0$, $x_0 = 0$ (Fig. 17). Then

it is easy to see that both solutions $\varphi_1 \equiv 0$, $\varphi_2 = (t/3)^3$ satisfy equation (1) and the condition (2). Of course, the function $\mathbf{v}$ is nondifferentiable, so that

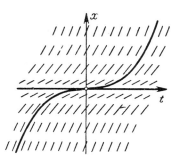

Fig. 17   An example of nonuniqueness.

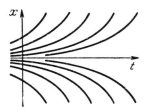

Fig. 18   Integral curves of the equation $\dot{x} = kx$.

this example does not contradict the theorem *as stated*. However the proof just given makes no use of the differentiability of **v** and goes through even in the case where the function **v** is merely continuous. Hence the proof cannot be correct as given. In fact, the uniqueness assertion 2) was proved only for the case $\mathbf{v}(x_0) \neq 0$, and we see that if the field **v** is only continuous (and not differentiable), then uniqueness may well fail for solutions satisfying the condition $\varphi(t_0) = x_0$ where x_0 is a singular point $(\mathbf{v}(x_0) = 0)$. It turns out, however, that differentiability of **v** guarantees uniqueness even in this case.

2.6. Example. Let $\mathbf{v}(x) = kx$, $U = \mathbf{R}$ (Fig. 18). Using (3) to solve the differential equation

$$\dot{x} = kx, \qquad k \neq 0 \tag{4}$$

of the form (1) subject to the condition (2), we get

$$t - t_0 = \int_{x_0}^{\varphi(t)} \frac{d\xi}{k\xi} = \frac{1}{k} \ln \frac{\varphi(t)}{x_0},$$

where φ is a solution such that $\varphi(t_0) = x_0 > 0$. Therefore

$$\varphi(t) = x_0 e^{k(t - t_0)} \tag{5}$$

for all t in a sufficiently small neighborhood of t_0.

Note that the right-hand side of (5) is defined on the whole t-axis, and represents an everywhere differentiable function satisfying the initial condition $\varphi(t_0) = x_0$ and the differential equation (4) for all t. In fact, it was precisely as the solution of equation (4) that Napier originally introduced the exponential function.

Problem 1. Prove that every solution φ of equation (4), satisfying the condition $\varphi(t_0) = x_0 > 0$, is given by formula (5) on the whole interval $a < t < b$ where it is defined.

Solution. We can argue, for example, as follows: Let T be the least upper bound of the set of numbers τ such that (5) holds for all t, $t_0 \leqslant t < \tau$. By hypothesis, $t_0 \leqslant T \leqslant b$. If $T < b$, formula (5) holds for $t = T$ because of the continuity of φ. But then it holds in some neighborhood of T (to prove this, repeat the argument leading to (5), replacing t_0 by T and x_0 by $\varphi(T)$ and noting that (5) implies $\varphi(T) > 0$). Thus $T = b$ and formula (5) is proved for $t_0 \leqslant t < b$. The case $a < t \leqslant t_0$ is treated similarly. Hence formula (5) gives all the solutions of (4) with $x_0 > 0$.

Comment. Thus the problems posed in Sec. 1.6 on radioactive decay and growth of bacterial colonies have been solved. In the first problem the amount of matter falls off exponentially with time. The amount of radioactive substance decreases to one half the amount initially present in a time $T = k^{-1} \ln 2$, called the *half-life* of the given substance. In the second problem, the number of bacteria grows exponentially with time, and doubles in time $T = k^{-1} \ln 2$ (as long as the food lasts). Formula (5) also contains the solution of many other problems (Fig. 19).

Problem 2. At what altitude is the density of the atmosphere one half its value at the earth's surface, assuming that the temperature is constant? (A cubic meter of air weighs ≈ 1250 gm at the earth's surface.)

Ans. $8 \ln 2 \approx 5.6$ km, the height of Mt. Elbrus.

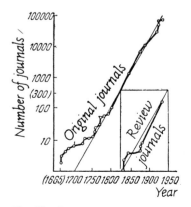

Fig. 19   Growth of the number of scientific journals (both original and review journals). From V. V. Nalimov and Z. M. Mulchenko, *Scientometry* (in Russian), Moscow (1969).

Problem 3. Prove that all the solutions of equation (4) satisfying the initial condition $\varphi(t_0) = x_0 < 0$ are also given by formula (5).

It should be noted that none of the functions (5) with $x_0 \neq 0$ vanishes for any value of t. Hence the unique solution of equation (4) such that $x_0 = 0$ is the stationary solution $x \equiv 0$. Thus *formula (5) accounts for all the solutions of the differential equation (4)*.

In particular, the uniqueness assertion of Theorem 2.3 is valid for equation (4). From this one can easily infer uniqueness for any equation (1) with a differentiable vector field **v** and for more general equations as well.

The reason for the failure of uniqueness in the case $\mathbf{v}(x) = x^{2/3}$ is that this field does not fall off fast enough as the point $x = 0$ is approached. Therefore the solution manages to arrive at the singular point in a finite time. An infinite time is required to reach the singular point in the case $\mathbf{v}(x) = kx$, since the integral curves approach each other exponentially. It is characteristic of any differential equation with a differentiable vector field **v** that its integral curves do not approach each other more rapidly than exponentially, thereby accounting for the uniqueness. In particular, the uniqueness proof in Theorem 2.3 is easily obtained by comparing the general equation (1) with a suitable equation of the form (4).

2.7. A comparison theorem. Let $\mathbf{v}_1, \mathbf{v}_2$ be real functions continuous on an interval U of the real axis such that $\mathbf{v}_1 < \mathbf{v}_2$, and let φ_1, φ_2 be solutions of the differential equations

$$\dot{x} = \mathbf{v}_1(x), \qquad \dot{x} = \mathbf{v}_2(x) \tag{6}$$

respectively, satisfying the same initial condition $\varphi_1(t_0) = \varphi_2(t_0) = x_0$ (Fig. 20), where φ_1, φ_2 are both defined on the interval $a < t < b$ $(-\infty \leqslant a < b \leqslant +\infty)$.

THEOREM. *The inequality*

$$\varphi_1(t) \leqslant \varphi_2(t) \tag{7}$$

holds for all $t \geqslant t_0$ in the interval (a, b).

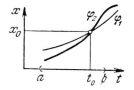

Fig. 20   The slope of φ_2 is greater than that of φ_1 at points with equal x, but not at points with equal t.

Proof. The inequality (7) is almost obvious ("the slower rider does not go further").† More exactly, let T be the least upper bound of the set of numbers τ such that (7) holds for all t, $t_0 \leqslant t < \tau$. By hypothesis, $t_0 \leqslant T \leqslant b$. If $T < b$, then $\varphi_1(T) = \varphi_2(T)$ by the continuity of φ_1, φ_2 and

$$\left.\frac{d\varphi_1}{dt}\right|_{t=T} < \left.\frac{d\varphi_2}{dt}\right|_{t=T}$$

by hypothesis, so that $\varphi_1 < \varphi_2$ at all points $t > T$ sufficiently near T. But then T cannot be the indicated least upper bound. This contradiction shows that $T = b$, as asserted. ∎

Remark. In the same way, it can be shown that $\varphi_1(t) \geqslant \varphi_2(t)$ for $t \leqslant t_0$.

2.8. Completion of the proof of Theorem 2.3. Let x_0 be a stationary point of a differentiable vector field $\mathbf{v}$, so that $\mathbf{v}(x_0) = 0$. Then, as we now show, the solution of equation (1) satisfying the initial condition (2) is unique, i.e., if φ is any solution of (1) such that $\varphi(t_0) = x_0$, then $\varphi(t) \equiv x_0$. There is no loss of generality in assuming that $x_0 = 0$. Since the field $\mathbf{v}$ is differentiable and $\mathbf{v}(0) = 0$, we have

$$|\mathbf{v}(x)| < k|x| \tag{8}$$

for sufficiently small $|x| \neq 0$, where $k > 0$ is a positive constant. The required uniqueness now follows from the fact that the integral curves of equation (4) other than $x = 0$, which are steeper near $x = 0$ than the integral curves of (1), cannot reach the line $x = 0$ in a finite time, as already noted in Sec. 2.6.

This can be proved more rigorously, for example, as follows: Let φ be a solution of (1) and (2) such that $\varphi(t_0) = 0$ (Fig. 21), and suppose $\varphi(t_1) > 0$, $t_1 > t_0$. Since φ is a continuous function, there exists an interval (t_2, t_3) with the following properties: 1) $\varphi(t_2) = 0$, 2) $\varphi(t_2) > 0$ for $t_2 < t \leqslant t_3$, 3) $x = \varphi(t)$ satisfies (3) for $t_2 < t \leqslant t_3$. In fact, for t_2 we can choose the greatest lower bound of the τ such that $\varphi(\tau) > 0$ for $\tau < t \leqslant t_1$ and for t_3 any point $t_3 > t_2$ sufficiently near t_2.

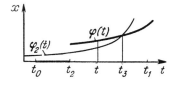

Fig. 21   The solution φ cannot vanish since it approaches zero more slowly than the exponential φ_2.

---

† Nevertheless we note that the rate of change of φ_1 *at a given instant* can be larger than the rate of change of φ_2 at the same instant (Fig. 20).

We now compare the solution $\varphi(t)$, $t_2 < t \leqslant t_3$ with the solution

$$\varphi_2(t) = \varphi(t_3)e^{k(t-t_3)}$$

of equation (4) subject to the initial condition $\varphi_2(t_3) = \varphi(t_3)$. Because of (8), the comparison theorem implies

$$\varphi(t) \geqslant \varphi(t_3)e^{k(t-t_3)}$$

for all $t_2 < t \leqslant t_3$, and hence

$$\varphi(t_2) \geqslant \varphi(t_3)e^{k(t_2-t_3)} > 0$$

by continuity. This contradicts $\varphi(t_2) = 0$ and shows that there is no t_1 such that $\varphi(t_1) > 0$, $t_1 > t_0$. The cases $t_1 < t_0$ and $\varphi(t_1) < 0$ are treated similarly. ∎

Problem 1. Prove the uniqueness by the method of Sec. 2.6, without making a comparison with equation (4). Prove that a sufficient condition for uniqueness is that the integral

$$\int_{x_0}^x \frac{d\xi}{\mathbf{v}(\xi)}$$

be divergent at x_0.

Problem 2. Prove the uniqueness assertion for the differential equation $\dot{x} = \mathbf{v}(x, t)$ where $\mathbf{v}$ is a differentiable function, assuming the existence of a solution $x = \varphi(t)$ satisfying the initial condition $\varphi(t_0) = x_0$.

Hint. Let $y = x - \varphi(t)$, and make a comparison with a suitable equation (4).

3. Phase Flows on the Line

Having just learned how to solve differential equations determined by a vector field on the line, we now see what our results mean in the language of phase flows.

3.1. One-parameter groups of linear transformations. We begin with the particularly simple equation

$$\dot{x} = kx, \qquad x \in \mathbf{R}. \tag{1}$$

As we know, the solution of (1) satisfying the initial condition $\varphi(0) = x_0$ is just

$$\varphi(t) = e^{kt}x_0.$$

We now define a "*t-advance mapping* g^t: $\mathbf{R} \to \mathbf{R}$," carrying the initial condition x_0 into the solution after time t:

$$g^t x_0 = e^{kt}x_0.$$

The family of mappings $\{g^t\}$ is called the *phase flow* associated with equation (1), or with the vector field $\mathbf{v} = kx$. Note that the mapping g^t is a linear transformation of the line, namely an expansion of the line e^{kt} times. For arbitrary real s and t we have

$$g^{s+t} = g^s g^t, \qquad g^0 x = x.$$

Moreover $g^t x$ is differentiable with respect to both t and x. It follows that *the phase flow* $\{g^t\}$ *is a one-parameter group of diffeomorphisms, where each diffeomorphism is a linear transformation of the line.* A one-parameter group of diffeomorphisms of a linear space, where each diffeomorphism is a linear transformation, will be called simply a *one-parameter group of linear transformations.*† Thus the phase flow $\{g^t\}$ associated with equation (1) is a one-parameter group of *linear* transformations, and the motions of points under the action of this phase flow are just the solutions of equation (1).

THEOREM. *Every one-parameter group* $\{g^t\}$ *of linear transformations of the line* **R** *is the phase flow of a differential equation of the form* (1), *so that*

$$g^t x = e^{kt} x$$

for some k.

Before proving the theorem, we make a remark of a general character.

3.2. The differential equation of a one-parameter group. Let $\{g^t\}$ be
a one-parameter group of diffeomorphisms of a domain U, and let **v** be the vector field of the phase velocity defined by the relation

$$\mathbf{v}(x) = \frac{d}{dt}\bigg|_{t=0} g^t x, \qquad x \in U.$$

THEOREM. *The motion of the phase point* $\varphi : \mathbf{R} \to U$, $\varphi(t) = g^t x$ *is a solution of the differential equation*

$$\dot{x} = \mathbf{v}(x). \tag{2}$$

Proof. We need only show that the velocity of motion of the phase point $g^t x$ at every instant of time t_0 coincides with the phase velocity at the point $g^{t_0} x$. This is obvious, since the transformations g^t form a group:

$$\frac{d}{dt}\bigg|_{t=t_0} g^t x = \frac{d}{d\tau}\bigg|_{\tau=0} g^{t_0+\tau} x = \frac{d}{d\tau}\bigg|_{\tau=0} g^\tau(g^{t_0} x) = \mathbf{v}(g^{t_0} x). \quad \blacksquare$$

3.3. The general form of a one-parameter group of linear transformations on the line. Let $\{g^t\}$ be a one-parameter group of *linear* transformations of a linear space L. Then the phase velocity $\mathbf{v}(x)$ depends on $x \in L$

---

† Note that *differentiability* with respect to t is implicit in the definition of a one-parameter group of linear transformations g^t.

linearly, since the derivative $(d/dt)|_{t=0}$ with respect to the parameter t of the function $g(t, x) = g^t x$ which is linear in x is itself linear in x.† In particular, if L is the real line $\mathbf{R}$, then every function linear in x is of the form $\mathbf{v}(x) = kx$ where $k = \mathbf{v}(1)$. Therefore the motion $\varphi(t) = g^t x$ is a solution of equation (2) with $\mathbf{v}(x) = kx$, i.e., a solution of equation (1). Since the unique solution φ of this equation satisfying the condition $\varphi(0) = x$ is of the form $g^t x = e^{kt} x$, the proof of Theorem 3.1 is now complete. ∎

*Problem 1. Prove that every continuous one-parameter group of linear transformations of the line is automatically differentiable.

Hint. Recall the definition of the exponential function for integral, rational, and irrational values of the argument.

Comment. Thus in defining a one-parameter group of linear transformations we could have replaced the requirement that the transformations g^t be differentiable with respect to t by the requirement that they be continuous in t.

*Problem 2. Find all one-parameter groups of linear transformations of the following linear spaces: a) $\mathbf{R}^2$ (the real plane); b) $\mathbf{C}^1$ (the complex line, i.e., the one-dimensional linear space over the field of complex numbers).

Hint. In Chap. 3 we will describe all one-parameter groups of linear transformations of the n-dimensional real and complex spaces $\mathbf{R}^n$ and $\mathbf{C}^n$.

3.4. A nonlinear example. Next we consider the more complicated differential equation

$$\dot{x} = \sin x, \qquad x \in \mathbf{R}.$$

Problem 1. Find the solution of this equation satisfying the initial condition $\varphi(0) = x_0$.

Here we can again define the t-advance mapping

$$g^t: \mathbf{R} \to \mathbf{R}, \qquad g^t x_0 = \varphi(t),$$

where $\varphi(t)$ is the solution satisfying the initial condition $\varphi(0) = x_0$. The mappings g^t form a one-parameter group of diffeomorphisms of the line, namely the phase flow associated with the given equation. The phase flow $\{g^t\}$ has fixed points $x = k\pi$, $k = 0, \pm 1, \ldots$, and the diffeomorphisms $g^t (t \neq 0)$ are nonlinear transformations of the line. The transformation g^t shifts every point x toward the nearest odd multiple of π if $t > 0$ and toward the nearest even multiple of π if $t < 0$ (Fig. 22).

Problem 2. Prove that the sequence of functions g^{t_i}, $t_i \to \infty$ converges, but not uniformly.

The above examples give rise to the hope that with every differential equation on the line

$$\dot{x} = \mathbf{v}(x), \qquad x \in \mathbf{R},$$

† Note that the linear nonhomogeneous function $f(x) = ax + b$ fails to be linear if $b \neq 0$.

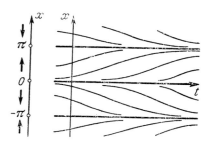

Fig. 22   Phase space and extended phase space of the equation $\dot{x} = \sin x$.

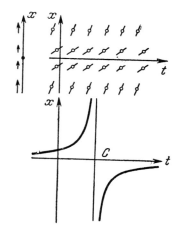

Fig. 23   Direction field and two solutions of the equation $\dot{x} = x^2$.

there is associated a one-parameter group $\{g^t\}$, $g^t x = \varphi(t)$ of diffeomorphisms of the line, where $\varphi(t)$ is the solution satisfying the condition $\varphi(0) = x$. As the next example shows, this hope is ill-founded.

3.5. Counterexample. Consider the differential equation

$$\dot{x} = x^2$$

characterizing "overly rapid growth" in the sense of Example 3 of Sec. 1.6 (Fig. 23). This equation has the solution

$$t - t_0 = \int_{x_0}^{\varphi(t)} \frac{d\xi}{\xi^2}$$

given by formula (3) of Sec. 2, often written in the form

$$\int dt = \int \frac{dx}{x^2},$$ (3)

$$t = -\frac{1}{x} + C, \qquad x = -\frac{1}{t - C}.$$

One must not think that the last formula is equivalent to (3) or that the function $x = -1/(t - C)$ is a solution. In fact, the domain of definition of the function $x = -1/(t - C)$ is not an interval but rather two intervals $t < C$ and $t > C$, so that the restriction of $x = -1/(t - C)$ to these intervals gives *two* solutions which are in no way related to each other (as long as we confine ourselves to the domain of real t, the only case considered in this book).

These considerations show that if the growth of a population is proportional to the number of pairs, then the size of the population becomes infinite in a finite time (whereas the usual law of growth is exponential). Physically this conclusion corresponds to the explosive nature of the process (of course, for t sufficiently near C, the idealization entailed in describing the process by the differential equation in question becomes inapplicable, so that the size of the population does not actually become infinite in a finite time). On the other hand, we see that *the formula for the t-advance mapping* $(g^t x_0 = \varphi(t)$ *where* $\varphi(t)$ *is the solution satisfying the initial condition* $\varphi(0) = x_0)$ *does not give a diffeomorphism* $g^t : \mathbf{R} \to \mathbf{R}$ *for any* $t \neq 0$.

Problem 1. Prove the italicized assertion.

3.6. Conditions for the existence of a phase flow.

The reason why $\{g^t\}$ in the preceding problem is not a one-parameter group of diffeomorphisms is not that differentiability fails or that the group property breaks down, but simply that the function g^t $(t \neq 0)$ *is not defined on the whole x-axis*, since some solutions manage to become infinite in a time not exceeding t (Fig. 24). However, if the solutions do not become infinite in a finite time, then the assertion made at the end of Sec. 3.4 is indeed valid.

Problem 1. Prove the assertion at the end of Sec. 3.4, assuming that the function $\mathbf{v}$ is differentiable and identically zero for sufficiently large $|x|$.

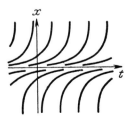

Fig. 24   Integral curves of the equation $\dot{x} = x^2$.

Hint. The solution is contained in the proof of a more general theorem, which asserts that every differentiable vector field on a *compact* manifold is the phase velocity field of a one-parameter group of diffeomorphisms (see Sec. 35).

Comment. Thus the possibility of the counterexample of Sec. 3.5 stems from the *noncompactness* of the line.

Problem 2. Prove the assertion of Sec. 3.5, assuming that $|\mathbf{v}(x)| < A|x| + B$ for all $x \in \mathbf{R}$, where A and B are positive constants.

Hint. Use the comparison Theorem 2.7.

4. Vector Fields and Phase Flows in the Plane

If the dimension of the phase space of a differential equation is greater than 1 (for example, equal to 2), then there is no general method for finding explicit solutions. However there are some special cases which can be reduced to one-dimensional problems.

4.1. Direct products. Consider two differential equations

$$\dot{x}_1 = \mathbf{v}_1(x_1), \qquad x_1 \in U_1, \tag{1}$$
$$\dot{x}_2 = \mathbf{v}_2(x_2), \qquad x_2 \in U_2, \tag{2}$$

determined by vector fields $\mathbf{v}_1$ and $\mathbf{v}_2$, differentiable in phase spaces U_1 and U_2, respectively.

Definition. By the *direct product of the differential equations* (1) *and* (2) is meant the differential equation whose phase space is the direct product of U_1 and U_2; this equation is determined by the vector field which is the "direct product" of the fields $\mathbf{v}_1$ and $\mathbf{v}_2$. Thus

$$\dot{x} = \mathbf{v}(x), \qquad x \in U, \tag{3}$$

where $U = U_1 \times U_2$, $x = (x_1, x_2)$, $\mathbf{v}(x) = (\mathbf{v}_1(x_1), \mathbf{v}_2(x_2))$.

In particular, if the phase spaces $U_1 \subset \mathbf{R}$ and $U_2 \subset \mathbf{R}$ are one-dimensional, then U is a domain in the plane (x_1, x_2) and the differential equation (3) is a system of two scalar differential equations of a special kind:

$$\begin{cases} \dot{x}_1 = \mathbf{v}_1(x_1), & x_1 \in U_1 \subset \mathbf{R}, \\ \dot{x}_2 = \mathbf{v}_2(x_2), & x_2 \in U_2 \subset \mathbf{R}. \end{cases} \tag{4}$$

The above definition immediately implies the following

THEOREM. *If φ is a solution of the direct product* (3) *of the differential equations* (1) *and* (2)*, then φ is a mapping $\varphi: I \to U$ of the form $\varphi(t) = (\varphi_1(t), \varphi_2(t))$, where φ_1 and φ_2 are solutions of equations* (1) *and* (2) *defined on one and the same interval I.*

In particular, if the phase spaces U_1 and U_2 are one-dimensional, we know how to solve each of the equations (1) and (2). Therefore we can also explicitly solve the system of two differential equations (4).

In fact, by Theorem 2.3, the solution φ satisfying the initial condition $\varphi(t_0) = x_0$ can be found in a neighborhood of the point $t = t_0$ from the relations

$$\int_{x_{10}}^{\varphi_1(t)} \frac{d\xi}{\mathbf{v}_1(\xi)} = t - t_0 = \int_{x_{20}}^{\varphi_2(t)} \frac{d\xi}{\mathbf{v}_2(\xi)}, \qquad x_0 = (x_{10}, x_{20})$$

if $\mathbf{v}_1(x_{10}) \neq 0$, $\mathbf{v}_{20}(x_{20}) \neq 0$. If $\mathbf{v}_1(x_{10}) = 0$, the first relation is replaced by $\varphi_1 \equiv x_{10}$, while if $\mathbf{v}_2(x_{20}) = 0$, the second relation is replaced by $\varphi_2 \equiv x_{20}$. Finally if $\mathbf{v}_1(x_{10}) = \mathbf{v}_2(x_{20}) = 0$, then x_0 is a singular point of the vector field $\mathbf{v}$ and an equilibrium position of the system (4), i.e., $\varphi(t) \equiv x_0$. ∎

4.2. Examples of direct products. Consider the following system of two differential equations:

$$\begin{cases} \dot{x}_1 = x_1, \\ \dot{x}_2 = kx_2. \end{cases}$$

Problem 1. Sketch the corresponding vector fields in the plane for $k = 0, \pm 1, \frac{1}{2}, 2$.

We have already solved each of these equations separately. Thus the solution φ satisfying the initial condition $\varphi(t_0) = x_0$ is of the form

$$\varphi_1 = x_{10}e^{t-t_0}, \qquad \varphi_2 = x_{20}e^{k(t-t_0)}. \tag{5}$$

Hence along every phase curve $x = \varphi(t)$ we have either $x_1 \equiv 0$ or

$$|x_2| = C|x_1|^k, \tag{6}$$

where C is a constant independent of t.

Problem 2. Is the curve in the phase plane (x_1, x_2) given by (6) a phase curve?

Ans. No.

The family of curves (6) where $C \in \mathbf{R}$ takes various forms depending on the value of the parameter k. If $k > 0$, we get a family of "generalized parabolas of exponent k,"† where the parabolas are tangent to the x_1-axis if $k > 1$ and to the x_2-axis if $k < 1$ (Figs. 25a and 25c). If $k = 1$, we get a family of straight lines going through the origin (Fig. 25b). The arrangement of phase curves shown in Fig. 25 is called a *node*. For $k < 0$ the curves are hyperbolas (Fig. 26),‡ forming a *saddle point* in a neighborhood of the origin. For $k = 0$ the curves turn into straight lines (Fig. 27).

It is clear from (5) that every phase curve lies entirely in one quadrant (or on one half of a coordinate axis, or possibly coincides with the origin which is a phase curve for all k). The arrows in the figures show the direction of

---

† The curves are actually parabolas only if $k = 2$ or $k = \frac{1}{2}$.
‡ The curves are actually hyperbolas only if $k = -1$.

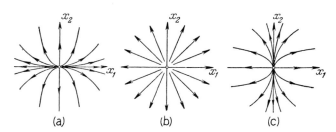

(a)   (b)   (c)

Fig. 25   Nodes: Phase curves of the system $\dot{x}_1 = x_1$, $\dot{x}_2 = kx_2$ for $k > 1$, $k = 1$, and $0 < k < 1$.

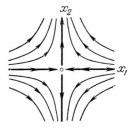

Fig. 26   A saddle point: Phase curves of the system $\dot{x}_1 = x_1$, $\dot{x}_2 = kx_2$, $k < 0$.

Fig. 27   Phase curves of the system $\dot{x}_1 = x_1$, $\dot{x}_2 = 0$.

motion of the point $\varphi(t)$ as t increases.

Problem 3. Prove that each of the parabolas $x_2 = x_1^2$ ($k = 2$) consists of three phase curves. Describe all the phase curves for the other values of k ($k > 1$, $k = 1$, $0 < k < 1$, $k = 0$, $k < 0$).

Comment. It is interesting to observe how one drawing goes into another as k changes continuously.

Problem 4. Draw the node corresponding to $k = 0.01$ and the saddle point corresponding to $k = -0.01$.

4.3. One-parameter groups of linear transformations of the plane.
Next we construct the phase flow associated with our system, defining the t-advance mapping g^t in the usual way, i.e., $g^t x = \varphi(t)$ where $\varphi(t)$ is the solution satisfying the initial condition $\varphi(0) = x$. It follows from (5) that g^t

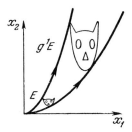

Fig. 28   Phase flow of the system $\dot{x}_1 = x_1$, $\dot{x}_2 = 2x_2$.

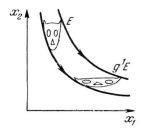

Fig. 29   Phase flow of the system $\dot{x}_1 = x_1$, $\dot{x}_2 = -x_2$. The transformations g^t are called hyperbolic rotations.

is a linear transformation of the plane, consisting of an e^t-fold expansion along the x_1-axis and an e^{kt}-fold expansion along the x_2-axis (an α-fold expansion is actually a contraction if $\alpha < 1$). The matrix of the transformation g^t has the diagonal form

$$\begin{pmatrix} e^t & 0 \\ 0 & e^{kt} \end{pmatrix}$$

in the system of coordinates x_1, x_2. The differentiability of $g^t x$ with respect to t and x is obvious. Thus the mappings g^t form a one-parameter group of linear transformations of the plane. The action of g^t, $t = 1$ on a set E is shown in Fig. 28 for the case $k = 2$ and in Fig. 29 for the case $k = -1$.

It should be noted that our one-parameter group of linear transformations g^t of the plane decomposes into the direct product of two one-parameter groups of linear transformations of the line (namely expansions along the x_1-axis and expansions along the x_2-axis).

Problem 1. Does every one-parameter group of linear transformations of the plane decompose in the same way?

Hint. Consider rotations through the angle t or shifts of the form $(x_1, x_2) \mapsto (x_1 + x_2 t, x_2)$.

5. Nonautonomous Equations

The simplest nonautonomous differential equation is of the form

$$\frac{dy}{dx} = f(x, y),$$

where the right-hand side depends on the independent variable x. We begin our discussion of such equations with the following example.

5.1. Equations with separable variables. Once again consider the direct product of two equations with one-dimensional phase spaces:

$$\begin{cases} \dot{x} = f(x), \\ \dot{y} = g(y). \end{cases} \tag{1}$$

Here $x \in U \subset \mathbf{R}$ is the coordinate in the first phase space and $y \in V \subset \mathbf{R}$ is the coordinate in the second phase space, while f and g are differentiable functions determining vector fields in U and in V. Suppose $f(x_0) \neq 0$, and consider the phase curve going through the point (x_0, y_0). Then, as we now show, this curve (Fig. 30) can be given by a curve of the form $y = F(x)$ in a neighborhood of the point (x_0, y_0).

Parametrically the phase curve is given by

$$x = \varphi_1(t), \qquad y = \varphi_2(t),$$

where $\varphi = (\varphi_1, \varphi_2)$ is the solution of the system (1) satisfying the condition $\varphi_1(t_0) = x_0$, $\varphi_2(t_0) = y_0$. Since $f(x_0) \neq 0$, we have

$$\left. \frac{d\varphi_1}{dt} \right|_{t=t_0} \neq 0.$$

By the implicit function theorem, the function ψ, $t = \psi(x)$ inverse to φ_1 is uniquely defined in a neighborhood of the point $x = x_0$. Let $F(x) = \varphi_2(\psi(x))$. Then the function F is defined, continuous, and differentiable in a

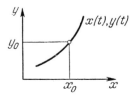

Fig. 30   A phase curve of the system (1) and an integral curve of equation (2).

neighborhood of the point $x = x_0$, and

$$\left.\frac{dF}{dx}\right|_{\xi} = \left.\frac{d\varphi_2}{dt}\right|_{t=\psi(\xi)} \left.\frac{d\psi}{dx}\right|_{\xi} = \frac{g(F(\xi))}{f(\xi)}, \qquad F(x_0) = y_0,$$

by the theorems on derivatives of composite and implicit functions. This is expressed concisely by saying that F is a solution of the differential equation

$$\frac{dy}{dx} = \frac{g(y)}{f(x)} \tag{2}$$

satisfying the initial condition $F(x_0) = y_0$. We call (2) an *equation with separable variables*.

THEOREM. *Let the functions f and g be defined and continuously differentiable in a neighborhood of the points $x = x_0, y = y_0$ respectively, where $f(x_0) \neq 0, g(y_0) \neq 0$. Then the solution F of equation (2) subject to the condition $F(x_0) = y_0$ exists and is unique† in a neighborhood of the point $x = x_0$, and satisfies the relation*

$$\int_{x_0}^{x} \frac{d\xi}{f(\xi)} = \int_{y_0}^{F(x)} \frac{d\eta}{g(\eta)}. \tag{3}$$

Proof. To construct a solution, consider the system (1). By Theorem 4.1, there exists a unique solution of (1) satisfying the initial condition $\varphi(t_0) = (x_0, y_0)$, given by the formula

$$\int_{x_0}^{x} \frac{d\xi}{f(\xi)} = t - t_0 = \int_{y_0}^{y} \frac{d\eta}{g(\eta)}$$

in some neighborhood of the point $t = t_0$. As shown above, the corresponding phase curve is the graph of the solution F of equation (2) subject to the initial condition $F(x_0) = y_0$. Hence the solution F exists and satisfies (3). The uniqueness is also a simple consequence of the relation between equations (1) and (2). ∎

Problem 1. Carry out the uniqueness proof.

Problem 2. Investigate the case where $g(y_0) = 0$.

Problem 3. Study the differential equation

$$\frac{dy}{dx} = k\frac{y}{x}$$

of the form (2) in the domain $x > 0, y > 0$.

Hint. The solution F satisfying the initial condition $F(x_0) = y_0$ is defined for all $x > 0$ and is given by the formula

$$F(x) = Cx^k, \qquad C = y_0 x_0^{-k}.$$

† In the sense that any two solutions coincide where they are defined.

See Figs. 25–27.

Problem 4. Draw graphs of the solutions of each of the differential equations

$$\frac{dy}{dx} = kx^\alpha y^\beta, \qquad \frac{dy}{dx} = \frac{\sin y}{\sin x}, \qquad \frac{dy}{dx} = \frac{\sin x}{\sin y}$$

in the domain where the right-hand side is defined.

5.2. Equations with variable coefficients. Let **v** be a differentiable mapping of a domain U in an $(n + 1)$-dimensional Euclidean space with coordinates $t, x_1, \ldots, x_n$ into an n-dimensional Euclidean space with co-ordinates $v_1, \ldots, v_n$. Such a mapping determines a *vector field* **v** *depending on the time t* and a corresponding *nonautonomous differential equation* or *equation with variable coefficients*

$$\dot{\mathbf{x}} = \mathbf{v}(t, \mathbf{x}) \tag{4}$$

or, in more detail,

$$\frac{dx_i}{dt} = v_i(t; x_1, \ldots, x_n), \qquad i = 1, \ldots, n.$$

Example 1. The differential equation (2) belongs to this class, with an obvious change of notation (here $n = 1$).

Definition. Let $\boldsymbol{\varphi}: I \to \mathbf{R}^n$ be a differentiable mapping defined on some interval I of the t-axis and taking values in the n-dimensional Euclidean space $\mathbf{R}^n$ with coordinates $x_1, \ldots, x_n$, such that the graph of $\boldsymbol{\varphi}$ lies in the domain U and

$$\left.\frac{d}{dt}\right|_{t=\tau} \boldsymbol{\varphi} = \mathbf{v}(\tau, \boldsymbol{\varphi}(\tau))$$

for every $\tau \in I$. Then $\boldsymbol{\varphi}$ is called a *solution* of the differential equation (4). If t is interpreted as the time and the space $\{\mathbf{x}\}$ is called phase space, then **v** can be regarded as a time-varying phase velocity field in phase space. In this language, a solution $\boldsymbol{\varphi}$ is the motion of a point in phase space such that the velocity of the point at every instant of time equals the value of the phase velocity vector at the point occupied by the moving point at the given instant.

Definition. A solution $\boldsymbol{\varphi}$ is said to *satisfy the initial condition* $\boldsymbol{\varphi}(t_0) = \mathbf{x}_0$ if the points t and $(t_0, \mathbf{x}_0)$ belong to I and U respectively, and if the value of $\boldsymbol{\varphi}$ at the point t_0 equals $\mathbf{x}_0$.

The solutions of a nonautonomous equation can be conveniently represented geometrically in the extended phase space $U \subset \mathbf{R}^1 \times \mathbf{R}^n$ (Fig. 31). Just as in the autonomous case, the right-hand side **v** determines a direction field in the domain U (if $n = 1$, **v** is the tangent of the angle of inclination

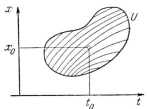

Fig. 31   Integral curves of the equation $\dot{x} = \mathbf{v}(x, t)$ in the extended phase space U.

with the positive t-axis).

Finding the solution satisfying the initial condition $\varphi(t_0) = \mathbf{x}_0$ is equivalent to drawing a curve through the point $(t_0, \mathbf{x}_0)$ of the domain U whose tangent at every point $(t, \mathbf{x} = \varphi(t))$ has a given direction. This curve (the graph of the solution) is called an *integral curve*.

Remark. Ordinarily the laws of nature do not vary with time, and equations like (4) with a time-dependent right-hand side arise most often in the following situation. Suppose we consider some part I of a physical system I + II. Then, although the law of evolution of the whole system does not vary with time, the influence of part II on part I may cause the law of evolution of part I to be time-dependent. For example, the influence of the moon on the earth produces tides, and this influence is expressed mathematically by the fact that the magnitude of the acceleration due to gravity figuring in the equation of motion of terrestrial objects becomes variable. In such situations, we say that the isolated part I is *nonautonomous*, which explains the term *nonautonomous system* as applied to (4). Of course, equations of the form (4) can occur in other situations as well, for example, in going from the pair of equations (1) to equation (2) with separated variables.

Problem 1. Find the solution φ of the differential equation

$$\dot{\mathbf{x}} = \mathbf{v}(t)$$

satisfying the initial condition $\varphi(t_0) = \mathbf{x}_0$.

Ans. It was to solve this problem that Newton introduced integration:

$$\varphi(t) = \mathbf{x}_0 + \int_{t_0}^{t} \mathbf{v}(t)\, dt.$$

Problem 2. Prove that the phase curves of the autonomous system

$$\dot{\mathbf{x}} = \mathbf{v}(\mathbf{x}), \qquad \mathbf{x} \in U \subset \mathbf{R}^n,$$

where $\mathbf{x} = (x_1, \ldots, x_n)$, $\mathbf{v} = (v_1, \ldots, v_n)$, $v_1 \neq 0$ are graphs of the solutions of the nonautonomous system

$$\frac{dx_i}{dx_1} = \frac{v_i(x)}{v_1(x)}, \qquad i = 1, \ldots, n-1,$$

and conversely.

5.3. Remarks on integration of differential equations. As shown above, the solutions of the simplest ordinary differential equations can be found by using the operation of integration. For this reason, the process of finding solutions of differential equations in general is sometimes called integration. There are a number of methods for integrating special kinds of differential equations, and lists of these equations and the corresponding methods can be found in the literature.[†] Anybody can enlarge the catalog of integrable differential equations by the simple device of making various substitutions in equations that have already been solved. Experts in integration of differential equations (like Jacobi) have in this way been very successful i n solving specific applied problems.

However, all these methods of integration have two fundamental shortcomings. In the first place, as shown by Liouville, *many differential equations cannot be solved in explicit form.* For example, even a simple equation like

$$\frac{dy}{dx} = y^2 - x$$

"cannot be solved by quadratures," i.e., the solution cannot be expressed as a finite combination of elementary functions or algebraic functions and integrals of such functions.[‡] Secondly, a complicated formula giving an explicit solution often turns out to be less useful than a simple approximate formula. For example, the equation $x^3 - 3x = 2a$ can be explicitly solved by Cardano's formula:

$$x = \sqrt[3]{a + \sqrt{a^2 - 1}} + \sqrt[3]{a - \sqrt{a^2 - 1}}.$$

However if we want to solve the equation for $a = 0.01$, it is useful to note that it has the root $x \approx -\frac{2}{3}a$ for small a, a fact which is hardly obvious from Cardano's formula. In just the same way, the pendulum equation $\ddot{x} + \sin x = 0$ can be solved in explicit form by using (elliptic) integrals, but most problems involving the behavior of a pendulum are more easily solved by starting from the approximate equation $\ddot{x} + x = 0$ for small oscillations and from qualitative considerations which do not involve an explicit formula (see Sec. 12).

† See e.g., A. F. Filippov, *Collection of Problems on Differential Equations* (in Russian), Moscow (1961) and E. Kamke, *Differential Equations, Methods of Solution and Solutions, I. Ordinary Differential Equations* (in German), Leipzig (1956), the latter containing some 1.6×10^3 equations.
‡ The proof of this fact resembles the proof of the nonsolvability of equations of degree 5 in terms of radicals (Ruffini-Abel-Galois), and is deduced from the nonsolvability of a certain group. Unlike ordinary Galois theory, we are concerned here with a nonsolvable Lie group rather than a nonsolvable finite group. The branch of mathematics dealing with these problems is called *differential algebra*.

Equations susceptible to exact solution are often useful as examples, since they sometimes exhibit behavior which occurs in more complicated cases as well. For example, this is true of so-called "self-similar solutions" of a number of equations of mathematical physics. Moreover, finding an exactly solvable problem always opens the possibility of solving neighboring problems approximately, by perturbation theory, say (see Sec. 9). However it is dangerous to extend results obtained by studying an exactly solvable problem to neighboring problems of a general form. In fact, an exactly integrable equation is often integrable precisely because its solutions are more simply behaved than those of neighboring nonintegrable problems.

6. The Tangent Space

In investigating various kinds of mathematical objects, it is always important to examine how the objects behave under mappings. A key role is played in the study of ordinary differential equations by changes of variables, i.e., by choice of a suitable coordinate system. Thus we must explain how the form of a differential equation changes under a differentiable mapping, and since a differential equation is specified by a vector field, the concepts of vector field and velocity vector must be analyzed.

Suppose we think of the velocity vector naively as an arrow made up of spatial points. Then under a mapping the arrow becomes curved and is no longer a vector. Below we will define a linear space whose elements are velocity vectors of curves going through a given point x of a domain U. This linear space is called the *tangent space* to U at the point x and is denoted by TU_x. Let $f: U \to V$ be a differentiable mapping. Then we will also define a linear mapping of tangent spaces

$$f_*|_x: TU_x \to TV_{f(x)},$$

called the *derivative* of the mapping f at the point x.

All the theorems in this section are essentially contained in a course on analysis, the only novelty being the fact that our terminology is more geometrical.

6.1. Definition of the tangent vector. Let U be a domain in n-dimensional Euclidean space with coordinates $x_i: U \to \mathbf{R}$, $i = 1, \ldots, n$, and let $\varphi: I \to U$ be a differentiable mapping of an open interval of the t-axis into U such that $\varphi(0) = x \in U$. Then we say that *the curve φ leaves the point x.*†

The *velocity vector of the curve φ at the point x in the system of coordinates x_i* is specified by its *components*

$$v_i = \frac{d}{dt}\bigg|_{t=0} (x_i \circ \varphi), \qquad i = 1, \ldots, n, \tag{1}$$

---

† More exactly, φ leaves the point x at the time $t = 0$. Of course, $t = 0$ can be replaced by $t = t_0$ by making appropriate changes in all the formulas.

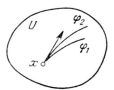

Fig. 32   Tangent curves.

where $(x_i \circ \varphi)(t) = x_i(\varphi(t))$ is the composite mapping $I \xrightarrow{\varphi} U \xrightarrow{x_i} \mathbf{R}$. The notation $v_i = \dot{x}_i|_{t=0}$ is also useful.

Definition. Two curves φ_1, $\varphi_2 \colon I \to U$ (Fig. 32) leaving the same point $x = \varphi_1(0) = \varphi_2(0)$ are said to be *tangent* (to each other) if the distance between the points $\varphi_1(t)$ and $\varphi_2(t)$ is $o(t)$, $t \to 0$.†

Problem 1. Prove that two curves are tangent at a point x if and only if their velocity vectors at the point x are the same.

The set of all tangent vectors of curves leaving x is an n-dimensional real linear space (with addition and multiplication by numbers being carried out component by component), called the *tangent space.*

Note that the coordinate system plays a role in this definition, and the resulting space seems at first glance to depend on the coordinate system. Thus we would now like to give an invariant definition of the velocity vector and the tangent space, which does not depend on the system of coordinates.

Definition. A system of coordinates $y_i \colon U \to \mathbf{R}$, $i = 1, \ldots, n$ in a domain U of Euclidean space $\mathbf{R}^n$ is said to be *admissible* if the mapping

$$y \colon U \to \mathbf{R}^n, \qquad y(x) = y_1(x)\mathbf{e}_1 + \cdots + y_n(x)\mathbf{e}_n$$

(with basis vectors $\mathbf{e}_i$ in $\mathbf{R}^n$) is a diffeomorphism.

Problem 2. Prove that the curves $y \circ \varphi_1$ and $y \circ \varphi_2$ leaving the point $y(x)$ are tangent if and only if the curves φ_1 and φ_2 leaving the point x are tangent (Fig. 33), so that *tangency of curves is a geometric concept, independent of the coordinate system.*

Definition. By the *velocity vector* $\mathbf{v}$ of a curve $\varphi \colon I \to U$ leaving a point $x \in U$ is meant the class of equivalent curves leaving x and tangent to φ (Fig. 34); in symbols,

$$\mathbf{v} = \dot{\varphi}(0), \qquad \mathbf{v} = \frac{d\varphi}{dt}\bigg|_{t=0}.$$

† Warning: The *ranges* of the mappings φ_1 and φ_2 can be lines perpendicular at x, say.

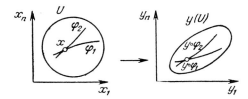

Fig. 33   Preservation of tangency under a diffeomorphism.

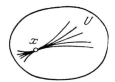

Fig. 34   Class of curves tangent at a point x.

Problem 3. Prove that tangency is an equivalence relation, i.e., that 1) $\xi \sim \xi$, 2) $\xi \sim \eta \Rightarrow \eta \sim \xi$, 3) $\xi \sim \eta \sim \zeta \Rightarrow \xi \sim \zeta$, where $\sim$ means "is tangent to, at x."

Remark. The coordinate system plays no role in our definition of the velocity vector, but *the class of admissible coordinate systems in U* does play a role. This class is called a *differentiable structure* in U. Without specifying a differentiable structure in U, one cannot define the concepts of tangency of curves or of the velocity vector of a curve φ.

6.2. Definition of the tangent space.

Definition. By the *tangent space to a domain U at a point x* is meant the set of all velocity vectors of the curves leaving x (Fig. 35). The elements of this set are called *tangent vectors*. The tangent space to U at the point x is denoted by TU_x (T for "tangent").†

Let $x_i: U \to \mathbf{R}, i = 1, \ldots, n$ be an admissible system of coordinates in U. Then the velocity vector of a curve $\varphi: I \to U$ leaving the point $x \in U$ has well-defined components $v_i \in \mathbf{R}, i = 1, \ldots, n$, given by formula (1) (see Problem 1). Thus the system of coordinates x_i determines a mapping $X: TU_x \to \mathbf{R}^n$ of the tangent space to U at the point x into the n-dimensional

† If the reader is accustomed to regard the velocity vector of a curve as lying in the same space as the curve itself, then the distinction between a tangent space to a linear space and the linear space itself may lead to certain psychological difficulties. In this case, it is helpful to repeat the preceding considerations with U thought of as the surface of a sphere. Then TU_x is the ordinary tangent plane.

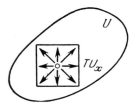

Fig. 35   The tangent space to a domain U at a point x.

real space $\mathbf{R}^n$ of vectors $(v_1, \ldots, v_n)$; the mapping X associates the numbers $v_1, \ldots, v_n$ with the velocity vector of the curve φ.

THEOREM 1. *The mapping $X: TU \to \mathbf{R}^n$ given by formula (1) is a one-to-one mapping of TU_x onto $\mathbf{R}^n$.*

Proof. According to Problem 1, the tangent vector (i.e., the class $\{\varphi\}$ of curves $\varphi: I \to U$ which are tangent to each other) is uniquely defined by the components of the velocity vector in the system of coordinates x_i. It remains to show that every vector $(v_1, \ldots, v_n) \in \mathbf{R}^n$ is the velocity vector of some curve. To show this, we need only choose the curve φ determined by the conditions $(x_i \circ \varphi)(t) = x_i(x) + v_i t$. ∎

Thus in a fixed coordinate system our abstract definitions of the tangent vector and the tangent space coincide with the naive definitions based on visualizing little arrows in the Euclidean space containing U.

So far our tangent space TU_x is simply a set which is not endowed with any further structure. We now equip TU_x with the structure of a real linear space. Fixing a system of coordinates x_i, we can add tangent vectors and multiply them by numbers by using the preceding theorem to identify them with arrows $(v_1, \ldots, v_n)$. It turns out that the resulting operations are independent of our choice of admissible coordinate system.

Definition. Let $\xi \in TU_x$, $\eta \in TU_x$, $\lambda \in \mathbf{R}$. Then the linear combination $\xi + \lambda\eta \in TU_x$ is defined as

$$\xi + \lambda\eta = X^{-1}(X\xi + \lambda X\eta)$$

in terms of the one-to-one mapping $X: TU_x \to \mathbf{R}^n$ determined by the admissible system of coordinates x_i. In other words, we carry over into TU_x the linear structure of $\mathbf{R}^n$, identifying these sets with the help of the one-to-one mapping X.

THEOREM 2. *The linear combination $\xi + \lambda\eta$ is independent of the admissible coordinate system figuring in its definition, and depends only on ξ, η, and λ.*

Proof. Let $y_i \colon U \to \mathbf{R}$, $i = 1, \ldots, n$ be another admissible system of coordinates, and let $Y \colon TU_x \to \mathbf{R}^n$ be the corresponding mapping of the tangent space to U at the point x into the n-dimensional real space $\mathbf{R}^n$ of vectors $(w_1, \ldots, w_n)$. The mapping Y associates the numbers

$$w_i = \left. \frac{d}{dt} \right|_{t=0} (y_i \circ \varphi), \qquad i = 1, \ldots, n \tag{2}$$

with the class of the curve φ, and is one-to-one by Theorem 1. We must show that the mapping $YX^{-1} \colon \mathbf{R}^n \to \mathbf{R}^n$ is an isomorphism of linear spaces. It is already known that this mapping is one-to-one. Let $\varphi \colon I \to U$ be a curve whose velocity vector in the system of coordinates x_i has components $\dot{x}_i$. We now find the components $\dot{y}_i$ of the velocity vector of this curve in the system of coordinates y_i. The coordinates y_i can be expressed in terms of the coordinates x_i as functions $y_i (x_1, \ldots, x_n)$. By the rule for differentiation of a composite function, we have

$$\dot{y}_i|_0 = \sum_{j=1}^n \left. \frac{\partial y_i}{\partial x_j} \right|_x \dot{x}_j|_0,$$

or, more concisely,

$$\dot{\mathbf{y}} = \frac{\partial y}{\partial x} \dot{\mathbf{x}}. \tag{3}$$

Equation (3) gives the explicit form of the mapping YX^{-1}, and this mapping is a linear transformation. Thus the operations introduced above indeed equip TU_x with the structure of a real n-dimensional linear space *independently of the choice of admissible coordinate system.* ∎

Remark. The coordinates $\dot{x}_i$ and $\dot{y}_j$ are fixed in the domain space $\mathbf{R}^n = \{\dot{\mathbf{x}}\}$ and the range space $\mathbf{R}^n = \{\dot{\mathbf{y}}\}$. According to (3), the matrix of the mapping YX^{-1} in these coordinate systems is just the Jacobian matrix $(\partial y / \partial x)$.

6.3. The derivative of a mapping. Let $f \colon U \to V$ be a differentiable mapping of a domain of n-dimensional Euclidean space with coordinates $x_i \colon U \to \mathbf{R}, i = 1, \ldots, n$ into a domain V of m-dimensional Euclidean space with coordinates $y_j \colon V \to \mathbf{R}, j = 1, \ldots, m$. Let x be a point of the domain U, and let $y = f(x) \in V$ be its image (Fig. 36).

Definition. By the *derivative of the mapping f at the point x* is meant the mapping

$$f_*|_x \colon TU_x \to TV_{f(x)}$$

of the tangent space to U at the point x into the tangent space to V at the point $f(x)$ which carries the velocity vector ξ leaving the point x of the curve

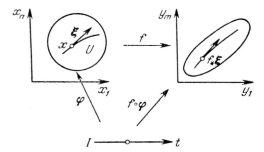

Fig. 36   Definition of the derivative of a mapping f at a point x.

$\varphi: I \to U$ into the velocity vector leaving the point $f(x)$ of the curve $f \circ \varphi: I \to V$, i.e.,

$$f_*|_x \left(\frac{d\varphi}{dt} \bigg|_{t=0} \right) = \frac{d}{dt} \bigg|_{t=0} (f \circ \varphi). \tag{4}$$

THEOREM. *Formula* (4) *defines a linear mapping* $f_*|_x$ *of the tangent space* TU_x *into the tangent space* $TV_{f(x)}$.

Proof. We must verify first that the right-hand side of (4) is independent of the choice of the representative φ of the class of tangent curves at x, and secondly that the mapping $f_*|_x$ is linear. Let $\dot{x}_i$ denote the components of the velocity vector $\dot{\mathbf{x}}$ of the curve φ at the point x, and $\dot{y}_j$ the components of the velocity vector $\dot{\mathbf{y}}$ of the curve $f \circ \varphi$ at the point $f(x)$. By the rule for differentiation of a composite function, we have

$$\dot{y}_j = \sum_{i=1}^{n} \frac{\partial y_j}{\partial x_i} \dot{x}_i, \tag{5}$$

where $y_j(x_1, \ldots, x_n), j = 1, \ldots, m$ are the functions specifying the mapping f in the coordinates x_i, y_j. But both assertions of the theorem are contained in (5). ∎

In addition, (5) implies the following

Remark. Suppose that in TU_x and $TV_{f(x)}$ we introduce the components $\dot{x}_i$, $\dot{y}_j$ of the tangent vectors in the coordinate systems x_i, y_j respectively. Then the matrix of the linear mapping $f_*|_x: TU_x \to TV_{f(x)}$ is the Jacobian matrix $(\partial y / \partial x)$. It should be emphasized that *the mapping* $f_*|_x$ *is independent of the coordinate system*, the coordinates being used only to prove the theorem.

Problem 1. Find the derivative at $x = 0$ of the mapping $f: \mathbf{R} \to \mathbf{R}$ given by the formula $y = x^2$.

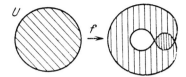

Fig. 37   A mapping which is a diffeomorphism in a neighborhood of every point may not be one-to-one.

Ans. $f_*|_0$ is the mapping of the line $T\mathbf{R}_0$ into the line $T\mathbf{R}_0$ carrying the whole line into 0.

Problem 2. Let $f: U \to V$, $g: V \to W$ be differentiable mappings. Prove that the composite mapping $h = g \circ f: U \to W$ is differentiable and that its derivative at the point x equals

$$h_*|_x = g_*|_{f(x)} \circ f_*|_x.$$

Problem 3. Let $f: U \to V$ be a diffeomorphism. Prove that the mapping $f_*|_x: TU_x \to TV_{f(x)}$ is an isomorphism of linear spaces. Give an example showing that *the converse is false* (see Fig. 37).

Problem 4. Let $f: \mathbf{R}^2 \to \mathbf{R}^2$ be the mapping given by the formula $(x_1 + ix_2)^2 = y_1 + iy_2$, $i = \sqrt{-1}$. Show that $f_*|_x$ $(x \neq 0)$ preserves angles (the Euclidean structures in $T\mathbf{R}^2_x$, $T\mathbf{R}^2_y$ are specified by quadratic forms $\dot{x}_1^2 + \dot{x}_2^2$ and $\dot{y}_1^2 + \dot{y}_2^2$ respectively).

6.4. The inverse function theorem. Let $f: U \to V$ be a differentiable mapping from one domain in Euclidean space to another, and let x_0 be a point of U.

THEOREM. *If the derivative*

$$f_*|_{x_0}: TU_{x_0} \to TV_{f(x_0)}$$

is an isomorphism of linear transformations, then there exists a neighborhood W of the point x_0 such that the restriction

$$f|_W: W \to f(W)$$

of f to W is a diffeomorphism.

Proof.[†] The dimensions of the tangent spaces TU_{x_0} and $TV_{f(x_0)}$, and hence the dimensions of the domains U and V, are the same. Let $x_1, \ldots, x_n$ be admissible coordinates in U, and $y_1, \ldots, y_n$ admissible coordinates in V. The mapping f is specified by functions $y_i = f_i(x_1, \ldots, x_n)$, $i = 1, \ldots, n$. Let

$$F_i(x_1, \ldots, x_n, y_1, \ldots, y_n) = y_i - f_i(x_1, \ldots, x_n).$$

By hypothesis, the determinant of the Jacobian matrix $(\partial f_i / \partial x_j)|_{x_0}$ is nonzero, i.e., the

---

† The inverse function theorem is easily deduced from the implicit function theorem, and vice versa. Here we derive the inverse function theorem from the implicit function theorem, since the latter always figures in courses on analysis while the former is usually left unstated. For a proof which is independent of the implicit function theorem, see, e.g., Sec. 31.9.

determinant of $(\partial F_i/\partial x_j)|_{x_0, f(x_0)}$ is nonzero. Applying the implicit function theorem to the system of functions F_i, $i = 1, \ldots, n$ in a neighborhood of the point $(x_0, f(x_0))$, we find that

1) In a sufficiently small neighborhood E of the point $y_0 = f(x_0)$ there exist n functions $x_i = \varphi_i(y_1, \ldots, y_n)$ such that $F(\varphi(y), y) \equiv 0$;

2) The system $F(x, y) = 0$, $y \in E$ has no other solutions x near x_0;

3) The values of the functions $\varphi_i(y)$ at the point y_0 equal the coordinates of the point x_0, and the φ_i are continuously differentiable the same number of times as the functions f_i in the neighborhood E of the point y_0 (Fig. 38). The functions φ_i determine a differentiable mapping φ of the neighborhood E of the point $y_0 = f(x_0)$ into a neighborhood of the point x_0 such that $f \circ \varphi$ is the identity mapping. Let $\varphi(E) = W$. Then the mappings $f|_W: W \to E$ and $\varphi: E \to W$ are mutually inverse differentiable mappings, and hence diffeomorphisms.   ∎

Problem 1. Prove that $\varphi(E)$ is a neighborhood of the point x_0, i.e., contains *all* points of the domain U which are sufficiently near the point x_0.

6.5. Action of a diffeomorphism on a vector field. Let U be a domain of Euclidean space, and let **v** be a vector field in U. If x is a point of the domain U, then $\mathbf{v}(x)$ is a tangent vector:

$$\mathbf{v}(x) \in TU_x.$$

Let $f: U \to V$ be a diffeomorphism.

Definition. By the *image of the vector field* **v** *under the diffeomorphism f* (Fig. 39) is meant the vector field $f_*\mathbf{v}$ whose vectors are obtained from the vectors $\mathbf{v}(x)$ by applying the derivative $f_*|_x$:

$$(f_*\mathbf{v})_{f(x)} = f_*|_x\mathbf{v}(x) \in TV_{f(x)}.$$

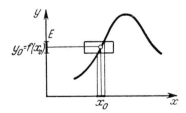

Fig. 38   The inverse function.

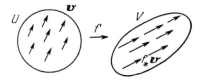

Fig. 39   Action of a diffeomorphism f on a vector field **v**.

Problem 1. Prove that if the field **v** is differentiable (i.e., is determined by r-fold continuously differentiable functions $v_i(x_1, \ldots, x_n)$ in the system of coordinates x_i), then the field $f_*\mathbf{v}$ is also differentiable (with the same r, if the diffeomorphism f is of class C^{r+1}).

Hint. See formula (5).

THEOREM. *Let $f: U \to V$ be a diffeomorphism. Then the differential equation*

$$\dot{x} = \mathbf{v}(x), \qquad x \in U \tag{6}$$

with phase space U determined by the vector field $\mathbf{v}$ is equivalent to the equation

$$\dot{y} = (f_*\mathbf{v})(y), \qquad y \in V \tag{7}$$

with phase space V determined by the vector field $f_*\mathbf{v}$, i.e., $\varphi: I \to U$ is a solution of (6) if and only if $f \circ \varphi: I \to V$ is a solution of (7).

Proof. Obvious. ∎

In other words, let $\varphi: I \to U$ be a solution of equation (6), and let $\tilde{\varphi}(\tau) = \varphi(t_0 + \tau)$. If $\varphi(t_0) = x_0$, then $\tilde{\varphi}$ leaves x_0 and $f \circ \tilde{\varphi}$ leaves $y_0 = f(x_0)$. It follows from the definition of f_* that

$$\left.\frac{d}{dt}\right|_{t=t_0} f \circ \varphi = \left.\frac{d}{d\tau}\right|_{\tau=0} f \circ \tilde{\varphi} = f_*|_{x_0} \left.\frac{d}{d\tau}\right|_{\tau=0} \tilde{\varphi}$$

$$= f_*|_{x_0} \left.\frac{d\varphi}{dt}\right|_{t=t_0} = (f_*\mathbf{v})(y_0).$$

Therefore $f \circ \varphi$ is a solution of equation (7). To complete the proof, we apply this result to the inverse diffeomorphism $f^{-1}: V \to U$.

6.6. Examples. The above theorem allows us to investigate and solve a great variety of differential equations. In fact, we need only take an equation that has already been solved and then apply a diffeomorphism, thereby solving the new equation as well.

Example 1. Consider the system

$$\begin{cases} \dot{x}_1 = x_2, \\ \dot{x}_2 = x_1, \end{cases} \tag{8}$$

determined by a vector field in the plane ($v_1 = x_2, v_2 = x_1$; Fig. 40). Let $f: \mathbf{R}^2 \to \mathbf{R}^2$ be the mapping carrying the point (x_1, x_2) into the point (y_1, y_2), where $y_1 = x_1 + x_2, y_2 = x_1 - x_2$. This linear mapping f is a diffeomorphism, and its derivative $f_*|_x$ has the matrix

$$\begin{pmatrix} 1 & 1 \\ 1 & -1 \end{pmatrix}.$$

Hence the new vector field $(f_*\mathbf{v})(y)$ has components $w_1 = y_1, w_2 = -y_2$, and our system is equivalent to the system .

$$\begin{cases} \dot{y}_1 = y_1, \\ \dot{y}_2 = -y_2. \end{cases}$$

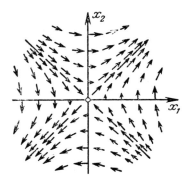

Fig. 40   The vector field $v_1 = x_2$, $v_2 = x_1$.

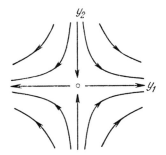

Fig. 41   The phase plane of the new system.

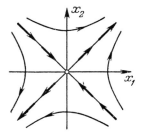

Fig. 42   The phase plane of the original system.

Fig. 43   A pendulum near its upper equilibrium position.

This system, which is a direct product of one-dimensional systems, has already been studied and solved. The system has a saddle point (Fig. 41) and a solution of the form

$$y_1 = y_1(0)e^t, \qquad y_2 = y_2(0)e^{-t}.$$

Using f^{-1} to return to the original system, we get a rotated saddle point (Fig. 42) and the solution

$$x_1(t) = x_1(0) \cosh t + x_2(0) \sinh t,$$
$$x_2(t) = x_1(0) \sinh t + x_2(0) \cosh t.$$

Remark. Let x be the angle of small deviation from the vertical of an inverted plane pendulum (Fig. 43). The equation of motion of the pendulum takes the form $\ddot{x} = x$ in an appropriate system of units.† Let $x = x_1$, $\dot{x}_1 = x_2$. Then the pendulum equation takes the form (8) for small deviations from the vertical equilibrium position.

Problem 1. To which motions of the pendulum do the various phase curves in Fig. 42 correspond?

Example 2. The equation $\ddot{x} = -x$ for small oscillations of a pendulum near its lower equilibrium position reduces to the system

$$\begin{cases} \dot{x}_1 = x_2, \\ \dot{x}_2 = -x_1 \end{cases} \tag{9}$$

if we write $x_1 = x$, $x_2 = \dot{x}$. The form of the vector field (Fig. 44) suggests the utility of polar coordinates

$$x_1 = r \cos \theta, \qquad x_2 = r \sin \theta.$$

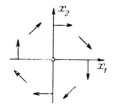

Fig. 44   The vector field of the pendulum equations (9).

† Actually $\ddot{x} = \sin x$, which can be approximated by $\ddot{x} = x$ for small x and $\dot{x}$. The difference in signs of the right-hand sides of the pendulum equation near its upper and lower equilibrium positions is explained as follows. In a neighborhood of the upper equilibrium position the moment of the force of gravity (the weight) moves the pendulum *in the direction of its inclination* and hence $\ddot{x} = +x$. In a neighborhood of the lower equilibrium position the moment of the force of gravity moves the pendulum *in the direction opposite to its inclination* and hence $\ddot{x} = -x$.

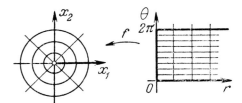

Fig. 45   Polar "coordinates."

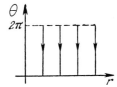

Fig. 46   Phase curves of the pendulum equations in polar coordinates.

These formulas give a differentiable mapping of the half-plane $r > 0$ onto the plane (x_1, x_2) minus the point 0 (Fig. 45). This mapping is not a diffeomorphism. However for the domain V we can choose the plane (x_1, x_2) minus any ray, say the ray $x_1 > 0$, while for the domain U we can choose the half-strip $0 < \theta < 2\pi$ in the half-plane $r > 0$. Then $f: U \to V$ is a diffeomorphism, and the system (9) in V is equivalent to a system in U, namely (Fig. 46)

$$\begin{cases} \dot{r} = 0, \\ \dot{\theta} = -1. \end{cases}$$

The solution of this system is of the form

$$r(t) = r(0), \qquad \theta(t) = \theta(0) - t,$$

and hence the original system (9) has the solution

$$x_1(t) = r_0 \cos(\theta_0 - t), \qquad x_2(t) = r_0 \sin(\theta_0 - t).$$

Problem 2. Verify that these formulas give all the solutions of the system (9) for all t, and not just for $(x_1, x_2) \in V$.

Problem 3. Prove that the phase curves are circles (Fig. 47), and that the t-advance mappings g^t form a one-parameter group of linear transformations of the plane, where g^t is a rotation through angle t with a matrix of the form

$$\begin{pmatrix} \cos t & \sin t \\ -\sin t & \cos t \end{pmatrix}.$$

Returning to the pendulum equation $\ddot{x} = -x$, we find that the pendulum

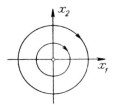

Fig. 47   Phase curves of the pendulum equations in rectangular coordinates.

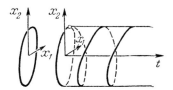

Fig. 48   Integral curves of the pendulum equations.

executes harmonic oscillations ($x = r_0 \cos(\theta_0 - t)$) whose period equals 2π and does not depend on the initial conditions.

Problem 4. What are the integral curves of the system (9)?

Ans. Helices of pitch $T = 2\pi$ with common axis $x_1 = x_2 = 0$, where the axis is also an integral curve (Fig. 48).

Example 3. Consider the system

$$\begin{cases} \dot{x}_1 = x_2 + x_1(1 - x_1^2 - x_2^2), \\ \dot{x}_2 = -x_1 + x_2(1 - x_1^2 - x_2^2), \end{cases} \tag{10}$$

obtained from the system

$$\begin{cases} \dot{r} = f(r), \\ \dot{\theta} = -1 \end{cases} \tag{11}$$

by going over to rectangular coordinates $x_1 = r\cos\theta$, $x_2 = r\sin\theta$. Actually, the system (11) is equivalent (with the usual stipulations involving the nonuniqueness of polar coordinates) to the system

$$\begin{cases} \dot{x}_1 = x_1 f(r)r^{-1} + x_2, \\ \dot{x}_2 = x_2 f(r)r^{-1} - x_1, \end{cases}$$

which reduces to (10) if $f(r) = r(1 - r^2)$.

Thus we must investigate the system (11) with $f(r) = r(1 - r^2)$. First we consider the integral curves of the equation $\dot{r} = f(r)$ in the half-plane (t, r), $r > 0$ (Fig. 49), noting that the vector field on the line $\mathbf{v} = f(r)$ has three singular points $r = \pm 1, 0$, where the field is directed toward the points $r = \pm 1$ and away from the point $r = 0$. The phase curves in the half-plane (r, θ), $r > 0$ are obtained by making a rotation (since $\theta = \theta_0 - t$). Returning to rectangular coordinates, we get the picture shown in Fig. 50. The curve $x_1 = x_2 = 0$ is the only singular point. The phase curves starting near this point move

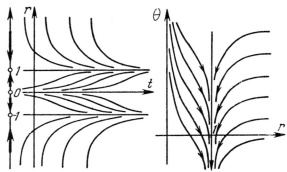

Fig. 49   Integral curves of the equation $\dot{r} = r(1 - r^2)$ and phase curves of the system (10) in polar coordinates.

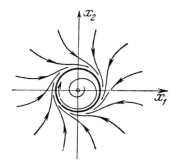

Fig. 50   Phase curves of the system (10). A limit cycle.

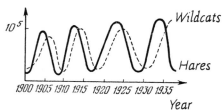

Fig. 51   Oscillations of the wildcat and hare populations in Canada.

away from it as t increases and wind around the circle $x_1^2 + x_2^2 = 1$ from the inside as $t \to +\infty$. This circle is itself a phase curve, called a *limit cycle*. However, if the initial point lies outside the disk $x_1^2 + x_2^2 \leqslant 1$, then the phase curve winds around the limit cycle from the outside as $t \to +\infty$ and goes off to infinity for negative t.

Limit cycles describe the stable periodic regimes of the motion of an autonomous system. For example, x_1 and x_2 might denote the deviations of the number of wildcats and the number of hares from their equilibrium values (the corresponding ecological equation is not exactly of the form (10), but has similar properties). Then the limit cycle corresponds to the periodic oscillations of the wildcat and hare populations, which are somewhat shifted in phase with respect to one another. This is actually observed in the field, with the oscillations in the number of wildcats lagging behind (Fig. 51).

Other examples of the occurrence of stable periodic oscillations under stationary external conditions are afforded by clocks, steam engines, electric bells, the human heart, vacuum tube oscillators generating radio waves, and variable stars of the Cepheid type; the operation of each of these mechanisms is described by a limit cycle in an appropriate phase space. However, it would be wrong to think that all oscillatory processes are described by limit cycles, and in fact much more complicated behavior of phase curves is possible in a multidimensional phase space. In this regard, we cite the precession of gyroscopes, the motion of planets and artificial satellites, including their rotations about their axes (the nonperiodicity of these motions is responsible for the complexity of the calendar and the difficulty in predicting tides), as well as the motion of charged particles in a magnetic field (responsible for the occurrence of the Aurora Borealis). See also Secs. 24 and 25.6.

2 Basic Theorems

In this chapter we formulate the basic results of the theory of ordinary differential equations, dealing with the existence and uniqueness of solutions and of first integrals, and with the dependence of solutions on initial data and parameters. We postpone the proofs until Chap. 4, confining ourselves at this point to a discussion of how the various results are related to one another.

7. The Vector Field near a Nonsingular Point

Consider the differential equation

$$\dot{\mathbf{x}} = \mathbf{v}(\mathbf{x}), \qquad \mathbf{x} \in U, \tag{1}$$

determined by a smooth vector field $\mathbf{v}$ in an n-dimensional phase space U. Let $\mathbf{x}_0 \in U$ be a nonsingular point of the vector field, so that $\mathbf{v}(\mathbf{x}_0) \neq 0$ (Fig. 52).

7.1. The basic theorem of the theory of ordinary differential equations. *The vector field $\mathbf{v}$ is diffeomorphic to a constant field $\mathbf{e}_1$ in every sufficiently small neighborhood of a nonsingular point. More exactly, there exists a neighborhood V of the point $\mathbf{x}_0$ and a diffeomorphism $f \colon V \to W$ of the neighborhood V onto a domain W of Euclidean space $\mathbf{R}^n$ (Fig. 53) such that $f_* \mathbf{v} = \mathbf{e}_1$ (where $\mathbf{e}_1$ is the first basis vector of $\mathbf{R}^n$). If $\mathbf{v}$ is a field of class C^r, $1 \leqq r \leqq \infty$, then f is a diffeomorphism of class C^r with the same r.*

Let $y_i \colon \mathbf{R}^n \to \mathbf{R}^1$, $i = 1, \ldots, n$ be rectangular coordinates in the Eu-

Fig. 52   Nonsingular point $\mathbf{x}_0$ of a vector field $\mathbf{v}$.

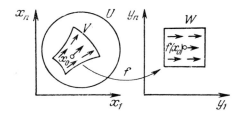

Fig. 53   Rectification of a vector field by a diffeomorphism f.

clidean space containing the domain W, so that the vector $\mathbf{e}_1$ has components $1, 0, \ldots, 0$. According to Sec. 6, the basic theorem can be formulated
as follows:

The differential equation (1), *considered in a sufficiently small neighborhood* V *of a
nonsingular point* $\mathbf{x}_0$, *is equivalent to the particularly simple equation*

$$\dot{\mathbf{y}} = \mathbf{e}_1, \qquad \mathbf{y} \in W, \tag{2}$$

i.e., to the system

$$\dot{y}_1 = 1, \qquad \dot{y}_2 = \cdots = \dot{y}_n = 0 \tag{3}$$

in the domain W.

The following is still another equivalent formulation of the basic theorem:
In a sufficiently small neighborhood V *of a nonsingular point* $\mathbf{x}_0$, *one can choose an
admissible coordinate system* $(y_1, \ldots, y_n)$ *such that equation* (1) *can be written in the
standard form* (3) *in these coordinates.*

The basic theorem is an assertion of the same character as the theorem of
linear algebra on reduction of quadratic forms or matrices of operators to
normal form. It gives an exhaustive description of the local behavior of a
vector field and of the differential equation (1) in a neighborhood of a nonsingular point $\mathbf{x}_0$, reducing everything to the case of the trivial equation (2).
The proof of the basic theorem will be given in Sec. 32.

7.2. Examples. The basic theorem might be called the *rectification theorem*,
since the phase curves and integral curves of equation (2) are straight lines.
Fig. 54 shows the level lines $y_i = \text{const}$ of the "rectifying coordinates" for
the pendulum equations.

Problem 1. Are the rectifying coordinates y_i uniquely defined? Prove that in the case
$n = 1$ the coordinate y is defined to within an affine transformation $y' = ay + b$.

Problem 2. Sketch level lines of rectifying coordinates for each of the following vector fields
in the domain U:
a) $\mathbf{v} = x_1\mathbf{e}_1 + 2x_2\mathbf{e}_2,$ $U = \{x_1, x_2 : x_1 > 0\}$;
b) $\mathbf{v} = \mathbf{e}_1 + \sin x_1\mathbf{e}_2,$ $U = \mathbf{R}^2$;
c) $\mathbf{v} = x_1\mathbf{e}_1 + (1 - x_1^2)\mathbf{e}_2,$ $U = \{x_1, x_2 : -1 < x_1 < 1\}$.

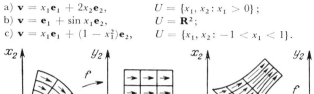

Fig. 54   Rectification of the pendulum equations.

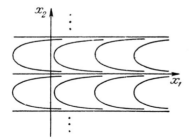

Fig. 55   A family of curves which is nonrectifiable in the whole plane.

Problem 3. Suppose that in $\mathbf{R}^3$ we are given a (differentiable) field of tangent planes $\mathbf{R}^2$. Can this field always be rectified (i.e., transformed into a field of parallel planes) in a neighborhood of a point with the help of a suitable diffeomorphism?

Hint. If the field of planes is rectifiable, then it is a field of planes tangent to the family of surfaces.

Ans. No. Consider for example the field of planes specified by the field of normals $x_2\mathbf{e}_1 + \mathbf{e}_3$ in $\mathbf{R}^3$. There does not exist a surface with this direction as the normal at each point.

Problem 4. Suppose a vector field $\mathbf{v}$ has no singular points in a domain U. Can one then rectify the field in the whole domain U, i.e., is the basic theorem true with $V = U$?

Hint. Construct the field in the plane whose phase curves are of the form shown in Fig. 55.

7.3. The existence theorem. The basic theorem immediately implies

COROLLARY 1. *There exists a solution of equation* (1) *satisfying the initial condition* $\boldsymbol{\varphi}(t_0) = \mathbf{x}_0$.

Proof. If $\mathbf{v}(\mathbf{x}_0) = 0$, let $\boldsymbol{\varphi}(t) \equiv \mathbf{x}_0$, while if $\mathbf{v}(\mathbf{x}_0) \neq 0$, then, by the basic theorem, equation (1) is equivalent to equation (2) in a neighborhood of the point $\mathbf{x}_0$. But (2) has a solution $\boldsymbol{\psi}$ (which?) satisfying the initial condition $\boldsymbol{\psi}(t_0) = \mathbf{y}_0 = f(\mathbf{x}_0)$. Hence equation (1), which is equivalent to (2), has a solution satisfying the initial condition $\boldsymbol{\varphi}(t_0) = \mathbf{x}_0$. ∎

7.4. The local uniqueness theorem. The basic theorem also immediately implies

COROLLARY 2. *Let* $\boldsymbol{\varphi}_1 : I_1 \to U$, $\boldsymbol{\varphi}_2 : I_2 \to U$ *be two solutions of equation* (1) *satisfying the same initial condition*

$$\boldsymbol{\varphi}_1(t_0) = \boldsymbol{\varphi}_2(t_0) = \mathbf{x}_0, \qquad \mathbf{v}(\mathbf{x}_0) \neq 0.$$

Then there exists an interval I_3 *containing* t_0 *on which* $\boldsymbol{\varphi}_1 \equiv \boldsymbol{\varphi}_2$.

Proof. This is obvious for equation (2). But equation (1) is equivalent to equation (2) in a sufficiently small neighborhood of the point $\mathbf{x}_0$. ∎

Remark. We will soon see that the restriction $\mathbf{v}(\mathbf{x}_0) \neq 0$ can be dropped. For $n = 1$ this has already been proved in Sec. 2.

7.5. Local phase flows. Let $\mathbf{v}$ be a vector field in the phase space U, and let $\mathbf{x}_0$ be a point of U.

Definition. By a *local phase flow* determined by the vector field $\mathbf{v}$ in a neighborhood of the point $\mathbf{x}_0$ we mean a triple (I, V_0, g), consisting of an interval $I = \{t \in \mathbf{R} : |t| < \varepsilon\}$ of the real t-axis, a neighborhood V_0 of the point $\mathbf{x}_0$, and a mapping $g : I \times V_0 \to U$, which satisfies the following three conditions:

1) For fixed $t \in I$ the mapping $g^t : V_0 \to U$ defined by $g^t\mathbf{x} = g(t, \mathbf{x})$ is a diffeomorphism;

2) For fixed $\mathbf{x} \in V_0$ the mapping $\boldsymbol{\varphi} : I \to U$ defined by $\boldsymbol{\varphi}(t_0) = g^t\mathbf{x}$ is a solution of equation (1) satisfying the initial condition $\boldsymbol{\varphi}(0) = \mathbf{x}$;

3) The group property $g^{s+t}\mathbf{x} = g^s(g^t\mathbf{x})$ holds for all $\mathbf{x}$, s, and t such that the right-hand side is defined, where for every point $\mathbf{x} \in V_0$ there exists a neighborhood $V, \mathbf{x} \in V \subset V_0$ and a number $\delta > 0$ such that the right-hand side is defined for $|s| < \delta, |t| < \delta$ and all $\mathbf{x} \in V$.

Example 1. Consider the vector field $\mathbf{v} = \mathbf{e}_1$ in a domain U of Euclidean space $\mathbf{R}^n$, and construct the following local phase flow in a neighborhood of a point $\mathbf{x}_0 \in U$. Start with a cube of side 4ε centered at $\mathbf{x}_0$ (Fig. 56). For sufficiently small ε this cube is entirely contained in U. Let V_0 denote the interior of the smaller cube of side 2ε (one half that of the original cube) and the same center, and let I be the interval $|t| < \varepsilon$. Then define the mapping g by the formula $g(t, \mathbf{x}) = \mathbf{x} + \mathbf{e}_1 t$.

Problem 1. Verify that conditions 1), 2), and 3) are satisfied for this example.

Another immediate consequence of the basic theorem is the following

COROLLARY 3. *The vector field* $\mathbf{v}$ *determines a local phase flow in a neighborhood of a nonsingular point* $\mathbf{x}_0$ $(\mathbf{v}(\mathbf{x}_0) \neq 0)$.

Proof. This has already been proved for equation (2). But, according to the

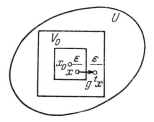

Fig. 56   Local phase flow of the equation $\dot{x}_1 = \mathbf{e}_1$.

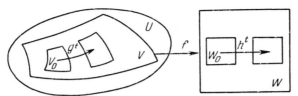

Fig. 57   The local phase flow (I, V_0, g) is obtained from the local phase flow (I, W_0, h) of the rectified equation by applying the diffeomorphism f^{-1}.

basic theorem, equation (1) is equivalent to equation (2) in a sufficiently small neighborhood of the point $\mathbf{x}_0$. ∎

In more detail, let (I, W_0, h) be a local phase flow of the field $\mathbf{e}_1$ in a neighborhood W of the point $\mathbf{y}_0 = f(\mathbf{x}_0)$, where $f: V \to W$ is the diffeomorphism figuring in the basic theorem. Then the desired phase flow is (I, V_0, g), where $V_0 = f^{-1}(W_0)$ and $g = f^{-1} \circ h^t \circ f$ (Fig. 57).

Remark 1. In particular, Corollary 3 asserts that

1) There exists an interval $|t| < \varepsilon$ on which a solution of equation (1) is defined satisfying any initial condition sufficiently near $\mathbf{x}_0$;

2) The value of this solution $\varphi(t)$ depends on t and $\mathbf{x}$ both continuously and differentiably (of class C^r if the field $\mathbf{v}$ is of class C^r).

Remark 2. We will soon see that the restriction $\mathbf{v}(\mathbf{x}_0) \neq 0$ can be dropped.

Problem 2. Prove that the value $\varphi(t)$ of the solution φ satisfying the initial condition $\varphi(t_0) = \mathbf{x}_0$ is differentiable with respect to t_0, $\mathbf{x}_0$, and t for sufficiently small $|t - t_0|$.

7.6. The theorem on continuous dependence and differentiability with respect to a parameter. The preceding theorem immediately implies

COROLLARY 4. *Let*

$$\dot{\mathbf{x}} = \mathbf{v}(\mathbf{x}, \boldsymbol{\alpha}), \qquad \mathbf{x} \in U \tag{1_α}$$

be a family of differential equations determined in the phase space U by vector fields $\mathbf{v}$ of class C^r and depending differentiably (of class C^r) on a parameter $\alpha \in A$, where A is a domain in Euclidean space. Suppose $\mathbf{v}(\mathbf{x}_0, \boldsymbol{\alpha}_0) \neq 0$. Then the value of the solution $\varphi(t)$ of equation (1_α) satisfying the initial condition $\varphi(0) = \mathbf{x}$ depends differentiably (of class C^r) on t, x, and α for sufficiently small $|t|$, $|\mathbf{x} - \mathbf{x}_0|$, and $|\alpha - \alpha_0|$.

Proof. Here a little ingenuity helps. Consider the vector field $(\mathbf{v}(\mathbf{x}, \boldsymbol{\alpha}), 0)$ in the direct product $U \times A$ (Fig. 58) and the corresponding system of equations

$$\begin{cases} \dot{\mathbf{x}} = \mathbf{v}(\mathbf{x}, \boldsymbol{\alpha}), \\ \dot{\boldsymbol{\alpha}} = 0. \end{cases} \tag{4}$$

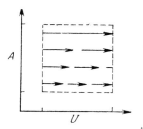

Fig. 58   The phase space of the extended system $\dot{\mathbf{x}} = \mathbf{v}(\mathbf{x}, \boldsymbol{\alpha})$, $\dot{\boldsymbol{\alpha}} = 0$.

By the preceding theorem, the solution of (4) depends differentiably on t, $\mathbf{x}$, and $\boldsymbol{\alpha}_0$ for sufficiently small $|t|$, $|\mathbf{x} - \mathbf{x}_0|$, and $|\boldsymbol{\alpha} - \boldsymbol{\alpha}_0|$. But the solution of (4) satisfying the initial condition $(\mathbf{x}, \boldsymbol{\alpha})$ is $(\boldsymbol{\varphi}, \boldsymbol{\alpha})$, where $\boldsymbol{\varphi}$ is the solution of equation (1_α) satisfying the initial condition $\boldsymbol{\varphi}(t_0) = \mathbf{x}$. Therefore $\boldsymbol{\varphi}(t)$ also depends differentiably on t, $\mathbf{x}$, and $\boldsymbol{\alpha}$. ∎

Remark. The condition $\mathbf{v}(\mathbf{x}_0, \boldsymbol{\alpha}_0) \neq 0$ can be dropped, as will be shown later.

7.7. The extension theorem. Let $\mathbf{v}$ be a vector field in a domain U, and let $\mathbf{x}_0$ be a point of U.

Definition. If there exists a solution $\boldsymbol{\varphi}$ of equation (1) satisfying the initial condition $\boldsymbol{\varphi}(t_0) = x_0$ defined for all $t \in \mathbf{R}$, we say that *the solution can be extended indefinitely.* If there exists a solution defined for all $t \geqslant t_0$ (or all $t \leqslant t_0$), we say that *the solution can be extended forward (or backward) indefinitely.*

Let Γ be a subset of the domain U. If there exists a solution $\boldsymbol{\varphi}$ of equation (1) satisfying the initial condition $\boldsymbol{\varphi}(t_0) = \mathbf{x}_0$ and defined on the interval $t_0 \leqslant t \leqslant T$ and if $\boldsymbol{\varphi}(T)$ belongs to Γ, we say that *the solution can be extended forward up to Γ. Extension backward up to Γ* is defined similarly.

Let F be a compact subset of a domain U containing a point $\mathbf{x}_0$, and let Γ denote the boundary of F (i.e., the set of points $\mathbf{x} \in F$ such that every neighborhood of $\mathbf{x}$ contains points of the complementary set $U \setminus F$). Suppose the vector field $\mathbf{v}$ in the domain U has no singular points. Then it is not hard to see that the basic theorem implies

COROLLARY 5. *The solution $\boldsymbol{\varphi}$ of equation (1) can be extended forward (or backward) either indefinitely or up to the boundary of F. This extension is unique in the sense that any two solutions satisfying the same initial condition coincide on the intersection of the intervals of definition.*

Proof. First we prove the uniqueness. Let T be the least upper bound of the set of numbers τ for which the solutions $\boldsymbol{\varphi}_1$ and $\boldsymbol{\varphi}_2$ coincide for all t in the interval $t_0 \leqslant t \leqslant \tau$ (Fig. 59). Suppose T is an interior point of both intervals of definition. Then $\boldsymbol{\varphi}_1(T) = \boldsymbol{\varphi}_2(T)$ because of the continuity of $\boldsymbol{\varphi}_1$ and $\boldsymbol{\varphi}_2$. By the local uniqueness theorem, $\boldsymbol{\varphi}_1$ coincides

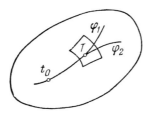

Fig. 59   The uniqueness of the extension follows from the local uniqueness theorem.

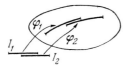

Fig. 60   Construction of the extension.

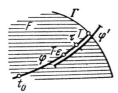

Fig. 61   Existence of the extension up to time T inclusive.

with φ_2 in a neighborhood of the point T, so that T cannot be the least upper bound. Hence T must be the end point of one of the intervals of definition, and the two solutions coincide on the part of the intersection of these intervals for which $t \geqslant t_0$. The case $t \leqslant t_0$ is treated similarly.

We now construct the extension. If the two solutions coincide on the intersection of the intervals of definition, then they can be combined to form a solution defined on the union of these intervals (Fig. 60). Let T be the least upper bound of the τ for which there exists a solution φ of equation (1) satisfying the initial condition $\varphi(t_0) = \mathbf{x}_0$ and also the condition $\varphi(t) \in F$ for all t, $t_0 \leqslant t \leqslant \tau$. By hypothesis, $t_0 \leqslant T \leqslant \infty$. If $T = \infty$, the solution can be extended indefinitely forward. Suppose $T < \infty$. Then, as we now show, there exists a solution φ defined for all t, $t_0 \leqslant t \leqslant T$, such that $\varphi(T) \in \Gamma$. In fact, it follows from Corollary 3 that every point $\mathbf{x}_0 \in U$ has a neighborhood $V_0(\mathbf{x}_0)$ and a corresponding number $\varepsilon(\mathbf{x}_0) > 0$ such that for all $\mathbf{x} \in V_0(\mathbf{x}_0)$ there exists a solution φ satisfying the initial condition $\varphi(t_0) = \mathbf{x}$ and defined for $|t - t_0| < \varepsilon$ (namely $\varphi = g^{t - t_0}\mathbf{x}$). Since F is compact, we can choose a finite covering of the set F from these neighborhoods of the points $\mathbf{x}_0 \in F$. Let $\varepsilon > 0$ be the smallest of the finite number of corresponding numbers $\varepsilon(\mathbf{x}_0)$. Since T is a least upper bound, there exists a τ between $T - \varepsilon$ and T such that $\varphi(t) \in F$ for all t in the interval $t_0 \leqslant t \leqslant \tau$. In particular $\varphi(\tau) \in F$, i.e., the point $\varphi(\tau)$ is covered by one of the neighborhoods of the finite covering. Hence there exists a solution φ' satisfying the initial condition $\varphi'(\tau) = \varphi(\tau)$ and defined for $|t - \tau| < \varepsilon$ (Fig. 61). By the uniqueness theorem, φ' coincides with φ on the whole intersection of the intervals of

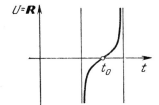

Fig. 62   The solutions of the equation $\dot{x} = x^2 + 1$ cannot be extended indefinitely either forward or backward.

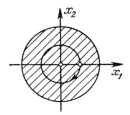

Fig. 63   The solution of the pendulum equations can be extended indefinitely, while remaining in the disk F.

definition. Hence we can use φ and φ' to construct a solution φ'' defined for $t_0 \leqslant t < \tau + \varepsilon$. In particular, $\varphi''(\tau)$ exists.

Finally, we show that $\varphi''(\theta) \in F$ if $t_0 \leqslant \theta < T$. In fact, every solution φ satisfying the initial condition $\varphi(t_0) = \mathbf{x}_0$ and defined for $t_0 \leqslant t \leqslant \theta$ must coincide with φ'' (uniqueness). If $\varphi''(\theta) = \varphi(\theta)$ did not belong to F, then T would not be the least upper bound of the set $\{\tau \colon \varphi(t) \in F \text{ for } t_0 \leqslant t \leqslant \tau\}$. Moreover, $\varphi''(T) \in \Gamma$. In fact $\varphi''(T) \in F$, being the limit of a sequence of points $\varphi''(\theta_i) \in F$, $\theta_i \to T$. On the other hand, every interval with T as its left-hand end point contains points t such that $\varphi''(t)$ does not belong to F, since otherwise all the points $\varphi''(t)$ would belong to F for all t in some neighborhood of T, and then T would not be the least upper bound. This proves the theorem for extension forward. The case $t < t_0$ is treated similarly. ∎[†]

Remark. We will soon see that the restriction $\mathbf{v}(\mathbf{x}) \neq 0$ for all $\mathbf{x} \in U$ can be dropped.

Example 1. Even in the case where U is the whole Euclidean space, the solution cannot always be extended indefinitely, e.g., when $n = 1$, $\mathbf{v}(x) = x^2 + 1$ (Fig. 62).

Example 2. Consider the pendulum equations $\dot{x}_1 = x_2, \dot{x}_2 = -x_1$. Let U be the plane (x_1, x_2) minus the origin of coordinates, and let F be the disk $|x_1|^2 + |x_2|^2 \leqslant 2$. Then the solution satisfying the initial condition $x_{1,0} = 1$, $x_{2,0} = 0$ can be extended indefinitely (Fig. 63).

[†] As always in proving obvious theorems, it is easier to carry out the proof of the extension theorem than to read through it.

Problem 1. For what initial conditions can the solution of equations with a limit cycle (Sec. 6.6, Example 3) be extended indefinitely?

Problem 2. Suppose every solution of equation (1) can be extended indefinitely both forward and backward. Let g^t denote the t-advance mapping (carrying every point $\mathbf{x}_0$ of the phase space U into the value $\boldsymbol{\varphi}(t)$ of the solution satisfying the initial condition $\boldsymbol{\varphi}(0) = \mathbf{x}_0$). Prove that $\{g^t\}$ is a one-parameter group of diffeomorphisms of U.

8. Applications to the Nonautonomous Case

We now consider the nonautonomous equation

$$\dot{\mathbf{x}} = \mathbf{v}(t, \mathbf{x}), \tag{1}$$

whose right-hand side is specified in a domain U of the *extended* phase space $\mathbf{R}^{n+1} = \mathbf{R} \times \mathbf{R}^n$, $t \in \mathbf{R}$, $\mathbf{x} \in \mathbf{R}^n$ (Fig. 64).

8.1. The basic theorem for the nonautonomous case. Let $(t_0, \mathbf{x}_0)$ be a point of the domain U. Then the basic theorem easily implies

COROLLARY 6. *There exists a neighborhood V of the point $(t_0, \mathbf{x}_0)$ in U and a diffeomorphism $f\colon V \to W$ of the neighborhood V onto a domain W in Euclidean $(n + 1)$-dimensional space with coordinates $t, y_1, \ldots, y_n$ such that equation (1) in V is equivalent to the particularly simple equation*

$$\frac{d\mathbf{y}}{dt} = 0, \qquad \mathbf{y} = (y_1, \ldots, y_n) \tag{2}$$

in W.

Thus the diffeomorphism f carries the point $(t, \mathbf{x})$ into the point $(t, \mathbf{y})$ while *leaving t unchanged*. The equivalence means that $\boldsymbol{\varphi}\colon I \to V$ is a solution of equation (1) if and only if $f \circ \boldsymbol{\varphi}\colon I \to W$ is a solution of equation (2).

The above corollary is equivalent to the basic theorem. A direct proof of the corollary will be given in Sec. 32.

Problem 1. Deduce Corollary 6 from the basic theorem.

Problem 2. Deduce the basic theorem from Corollary 6.

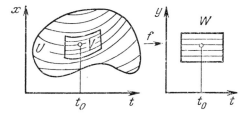

Fig. 64   Rectification of integral curves by a diffeomorphism f of extended phase space.

8.2. The existence theorem. Corollary 6 obviously implies

COROLLARY 7. *For sufficiently small* $|t - t_0|$ *there exists a solution of equation* (1) *satisfying the initial condition* $\boldsymbol{\varphi}(t_0) = \mathbf{x}_0 \in U.$

8.3. The uniqueness theorem. Another immediate consequence of Corollary 6 is given by

COROLLARY 8. *Any two solutions of equation* (1) *satisfying the same initial condition coincide on the intersection of the intervals on which they are defined.*

Proof. We need only note that this is obviously true for equation (2). ∎

Remark. Applying Corollary 8 to the case where $\mathbf{v}$ in (1) is independent of t, we see that the requirement $\mathbf{v}(\mathbf{x}_0) \neq 0$ can be dropped in Corollary 2 of Sec. 7.4.

8.4. The differentiability theorem. Let $\mathbf{v} = \mathbf{v}(t, \mathbf{x})$ be a vector field in a domain U of extended phase space. In the nonautonomous case, the t-advance mappings do not form a one-parameter group of transformations. However we can define "(t_1, t_2)-advance mappings" as follows:

Definition. By a *local family of transformations* $g_{t_1}^{t_2}$ determined by the field $\mathbf{v}(t, \mathbf{x})$ in a neighborhood of a point $(t_0, \mathbf{x}_0)$ is meant a triple (I, V_0, g) consisting of an interval I of the real axis containing t_0, a neighborhood V_0 of the point $\mathbf{x}_0$ in *phase* space, and a mapping $g: I \times I \times V_0 \to U$, such that
1) For fixed $t_1, t_2 \in I$ the mapping $g_{t_1}^{t_2}: (V_0 \times t_1) \to U$ defined by $g_{t_1}^{t_2}(\mathbf{x}, t_1) = g(t_2, t_1, \mathbf{x})$ is a diffeomorphism (on part of the plane $t = t_2$);
2) For fixed $\mathbf{x} \in V_0$, $t_1 \in I$ the mapping $\boldsymbol{\varphi}$ defined by $(\boldsymbol{\varphi}(t), t) = g(t, t_1, \mathbf{x})$ is a solution of equation (1) satisfying the initial condition $\boldsymbol{\varphi}(t_1) = \mathbf{x}$;
3) The property

$$g_{t_1}^{t_3}(\mathbf{x}, t_1) = g_{t_2}^{t_3} g_{t_1}^{t_2}(\mathbf{x}, t_1)$$

analogous to the group property (Fig. 65) holds for all $\mathbf{x}, t_1, t_2$, and t_3 for which the right-hand side is defined, where for every point $\mathbf{x} \in V_0$ there

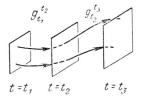

$$g_{t_1}^{t_2} \qquad g_{t_2}^{t_3}$$

$$t = t_1, \quad t = t_2 \qquad t = t_3$$

Fig. 65   A local family of transformations.

exists a neighborhood V, $\mathbf{x} \in V \subset V_0$ and a number $\delta > 0$ such that the right-hand side is defined for $|t_i - t_0| < \delta, i = 1, 2, 3$ and all $\mathbf{x} \in V$.

The basic theorem now immediately implies

COROLLARY 9. *The vector field* $\mathbf{v}(t, \mathbf{x})$ *determines a local family of transformations in a neighborhood of the point* $(t_0, \mathbf{x}_0)$.

Proof. Similar to that of Corollary 3. ∎

Remark. Identifying every plane $t = t_0$ in extended phase space with phase space, we can regard the mapping $g_{t_1}^{t_2}$ as a diffeomorphism of a domain of phase space into a domain of phase space. In the special case where equation (1) is autonomous and $\mathbf{v}(t, \mathbf{x}) = \mathbf{v}(\mathbf{x})$ is independent of t, the diffeomorphism $g_{t_1}^{t_2}$ depends only on the difference $t_2 - t_1$ and coincides with the $(t_2 - t_1)$-advance mapping $g^{t_2 - t_1}$. (This follows from the uniqueness theorem and from the fact that if $\mathbf{x} = \boldsymbol{\varphi}(t)$ is a solution of the autonomous equation, then so is $\mathbf{x} = \boldsymbol{\varphi}(t + C)$.)

Thus Corollary 9 contains Corollary 3 as a special case, *but without the restriction* $\mathbf{v}(\mathbf{x}) \neq 0$.

Problem 1. Prove that $g_{t_1}^{t_2}$ depends on just $t_2 - t_1$ if and only if $\mathbf{v}(t, \mathbf{x})$ is independent of t.

8.5. Dependence on a parameter. The following proposition is also an easy consequence of the basic theorem:

COROLLARY 10. *If* $\mathbf{v} = \mathbf{v}(t, \mathbf{x}, \boldsymbol{\alpha})$ *is a vector field depending* C^r-*differentiably on a parameter* $\boldsymbol{\alpha}$ *(as well as on t and* $\mathbf{x}$*), then the value* $\boldsymbol{\varphi}(t)$ *of the solution of the equation*

$$\dot{\mathbf{x}} = \mathbf{v}(t, \mathbf{x}, \boldsymbol{\alpha})$$

satisfying the initial condition $\boldsymbol{\varphi}(t_0) = \mathbf{x}_0$ *depends* C^r-*differentiably on* t_0, $\mathbf{x}_0$, $\boldsymbol{\alpha}$, *and t.*

Proof. Similar to that of Corollary 4. ∎

Note that Corollary 10 is applicable independently of whether or not $\mathbf{v}$ vanishes. Therefore Corollary 4 has now been proved without the restriction $\mathbf{v} \neq 0$.

8.6. The extension theorem. Let $\mathbf{v} = \mathbf{v}(t, \mathbf{x})$ be a vector field in a domain U of extended phase space, let $(t_0, \mathbf{x}_0)$ be a point of U, and let F be a compact set containing this point (Fig. 66). Then the basic theorem immediately implies

COROLLARY 11. *The solution* $\boldsymbol{\varphi}$ *of equation* (1) *satisfying the initial condition* $\boldsymbol{\varphi}(t_0) = \mathbf{x}_0$ *can be extended backward and forward up to the boundary of* F. *Any two*

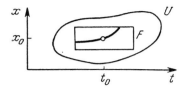

Fig. 66   Extension of a solution up to the boundary of a compact set F in extended phase space.

solutions satisfying the same initial condition coincide on the intersection of the intervals of definition.

Proof. Similar to that of Corollary 5.   ∎

Problem 1. Prove that Corollary 5 is valid even if the field **v** has singular points.

Problem 2. Suppose every solution of equation (1) can be extended indefinitely forward or backward. Prove that $g_{t_1}^{t_2}$ is a diffeomorphism of phase space onto itself.

Problem 3. Suppose, in addition, that the vector field **v** is periodic in time, so that $\mathbf{v}(t + T, \mathbf{x}) = \mathbf{v}(t, \mathbf{x})$ for all t and **x**. Prove that the diffeomorphisms $\{g_0^{nT}\}$ (n an integer) form a group, i.e.,

$$g_0^{nT} = A^n$$

where $A = g_0^T$. Which of the following two relations is true:

$$g_0^{nT+\tau} = A^n g_0^\tau, \qquad g^{nT+\tau} = g_0^\tau A^n\,?$$

9. Applications to Equations of Higher Order

By a differential equation *of order n* we mean an equation of the form

$$\frac{d^n x}{dt^n} = F\left(t, x, \frac{dx}{dt}, \frac{d^2 x}{dt^2}, \ldots, \frac{d^{n-1} x}{dt^{n-1}}\right),\tag{1}$$

where $F(u_0, u_1, \ldots, u_n)$ is a differentiable function (of class C^r, $r \geqslant 1$) defined in a domain U.

9.1. Equivalence of an equation of order n to a system of n first-order equations. By a *solution* of equation (1) we mean a C^n-mapping $\varphi: I \to \mathbf{R}$ of an interval $a < t < b$ (where $-\infty \leqslant a < b \leqslant +\infty$) of the real axis into the real axis such that

1) The point with coordinates

$$u_0 = \tau, \quad u_1 = \varphi(\tau), \quad u_2 = \left.\frac{d\varphi}{dt}\right|_{t=\tau}, \quad \ldots, \quad u_n = \left.\frac{d^{n-1}\varphi}{dt^{n-1}}\right|_{t=\tau}$$

belongs to the domain U for every $\tau \in I$;

2) For every $\tau \in I$,

$$\left.\frac{d^n\varphi}{dt^n}\right|_{t=\tau} = F\left(\tau, \varphi(\tau), \left.\frac{d\varphi}{dt}\right|_{t=\tau}, \ldots, \left.\frac{d^{n-1}\varphi}{dt^{n-1}}\right|_{t=\tau}\right).$$

For example, the functions $\varphi(t) = \sin t$ and $\varphi(t) = \cos t$ are both solutions of the equation

$$\frac{d^2x}{dt^2} = -x, \qquad x \in \mathbf{R}$$

for the small oscillations of a pendulum.

The phase space of the pendulum equation is the plane $(x, \dot{x})$, as in Sec. 1.6, Example 5. We now consider the question of the dimensionality of the phase space corresponding to the nth-order equation (1).

THEOREM. *Equation* (1) *is equivalent to the system*

$$\begin{cases} \dot{x}_1 = x_2, \\ \dot{x}_2 = x_3, \\ \ldots \\ \dot{x}_n = F(t, x_1, \ldots, x_n) \end{cases} \tag{2}$$

of n first-order equations in the sense that if φ is a solution of equation (1), *then the vector* $(\varphi, \dot{\varphi}, \ddot{\varphi}, \ldots, \varphi^{(n-1)})$ *made up of the derivatives of φ is a solution of the system* (2), *while if* $(\varphi_1, \ldots, \varphi_n)$ *is a solution of the system* (2), *then φ_1 is a solution of* (1).

Proof. Obvious.   ∎

Thus the phase space of any process described by a differential equation of order n is of dimension n. The whole course of the process φ is determined by specifying n numbers at time t_0, namely the values at t_0 of the derivatives of φ of order less than n.

Example 1. The pendulum equation is equivalent to the system

$$\begin{cases} \dot{x}_1 = x_2, \\ \dot{x}_2 = -x_1, \end{cases}$$

already investigated in Secs. 1.6 and 6.6.

Example 2. The equation $\ddot{x} = 0$ is equivalent to the system

$$\begin{cases} \dot{x}_1 = x_2, \\ \dot{x}_2 = 0, \end{cases}$$

whose solution is easily found to be $x_2(t) = x_2(0) = C$, $x_1(t) = x_1(0) + Ct$. Thus every solution of the equation $\ddot{x} = 0$ is a polynomial of the first degree in t.

Problem 1. Prove that the equation $d^n x/dt^n = 0$ is satisfied by all polynomials of degree less than n and only by these polynomials.

9.2. The existence and uniqueness theorems. Theorem 9.1 and Corollaries 7 and 8 to the basic theorem immediately imply

COROLLARY. *Given a point $u = (u_0, u_1, \ldots, u_n)$ of a domain U, the solution of equation* (1) *satisfying the initial conditions*

$$\varphi(u_0) = u_1, \quad \frac{d\varphi}{dt}\bigg|_{t=u_0} = u_2, \quad \ldots, \quad \frac{d^{n-1}\varphi}{dt^{n-1}}\bigg|_{t=u_0} = u_n \tag{3}$$

exists and is unique (in the sense that any two solutions satisfying (3) *coincide on the intersection of the intervals of definition).*

We can write the initial conditions (3) more concisely as

$$t = u_0, \quad x = u_1, \quad \dot{x} = u_2, \quad \ldots, \quad x^{(n-1)} = u_n.$$

Example 1. The solution of the pendulum equation $\ddot{x} = -x$ (Fig. 67) satisfying the initial conditions

$$t = 0, \qquad x = 0, \qquad \dot{x} = 0 \tag{I}$$

is $\varphi \equiv 0$. If the initial conditions are

$$t = 0, \qquad x = 0, \qquad \dot{x} = 1, \tag{II}$$

then $\varphi(t) = \sin t$, while if they are

$$t = 0, \qquad x = 1, \qquad \dot{x} = 0, \tag{III}$$

then $\varphi(t) = \cos t$.

Problem 1. Find the solutions of the equation $\ddot{x} = x$ of the inverted pendulum (Fig. 68), satisfying the initial conditions (I), (II), (III), and

$$t = 0, \qquad x = 1, \qquad \dot{x} = 1, \tag{IV}$$
$$t = 0, \qquad x = 1, \qquad \dot{x} = -1. \tag{V}$$

What are the motions of the pendulum corresponding to these solutions?

Counterexample 1. Consider the equation $2x = t^2\ddot{x}$ and the initial conditions $t = 0, x = 0, \dot{x} = 0$ (Fig. 69). Then many solutions satisfy these conditions, for example, $\varphi(t) \equiv 0$ and $\varphi(t) = t^2$. The point is that the equation in question is not of the form (1).

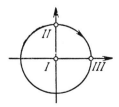

Fig. 67   Three special solutions of the pendulum equation.

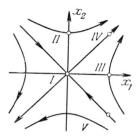

Fig. 68   Five special solutions of the equation of the inverted pendulum.

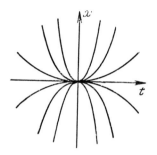

Fig. 69   Nonuniqueness of a solution satisfying the initial condition $x = \dot{x} = 0$.

9.3. The differentiability and extension theorems.

Problem 1. State and prove the theorem on continuous and differentiable dependence on the initial conditions and on parameters, and also the extension theorem, for the case of a differential equation of order n.

9.4. Systems of equations. By a *system of differential equations* we mean a system of equations of the form

$$\frac{d^{n_i} x_i}{dt^{n_i}} = F_i(t, x, \dots), \qquad i = 1, \dots, n \tag{4}$$

involving n unknown functions x_i, where the arguments of the functions F_i include the independent variable t, the unknown functions x_j, and the derivatives of the x_j of order less than n_j $(j = 1, \dots, n)$. Solutions of the system (4) are defined as in Sec. 9.1. It should be emphasized that a solution of the system is a vector function $(\varphi_1, \dots, \varphi_n)$ defined on an interval. Thus $(\varphi_1, \dots, \varphi_n)$ *is a single solution and not n solutions*, an observation which applies with equal force to systems of algebraic equations and systems of differential equations.

First of all, we explain the nature of the phase space corresponding to the system (4).

THEOREM. *The system* (4) *is equivalent to a system of*

$$N = \sum_{i=1}^{n} n_i$$

first-order differential equations. In other words, the dimension of the phase space of the system (4) *equals* N.

Proof. As in Sec. 9.1, introduce the derivatives of the x_i of order less than n_i as the coordinates in phase space.   ∎

For example, suppose $n = n_1 = n_2 = 2$. Then the system (4) is of the form

$$\ddot{x}_1 = F_1(t, x_1, x_2, \dot{x}_1, \dot{x}_2),$$
$$\ddot{x}_2 = F_2(t, x_1, x_2, \dot{x}_1, \dot{x}_2),$$

and is equivalent to the system of four equations

$$\dot{x}_1 = x_3, \qquad \dot{x}_2 = x_4, \qquad \dot{x}_3 = F_1(t, x), \qquad \dot{x}_4 = F_2(t, x),$$

where $x = (x_1, x_2, x_3, x_4)$.

Example 1. In mechanics the system of Newton's equations

$$m_i \ddot{q}_i = -\frac{\partial U}{\partial q_i}, \qquad i = 1, \dots, n \tag{5}$$

(where U is the potential energy and the $m_i > 0$ are masses) is equivalent to the system of $2n$ Hamilton equations

$$\dot{q}_i = \frac{\partial H}{\partial p_i}, \quad \dot{p}_i = -\frac{\partial H}{\partial q_i}, \quad i = 1, \dots, n,$$

where $p_i = m_i \dot{q}_i$,

$$T = \sum_{i=1}^{n} \frac{m_i \dot{q}_i^2}{2} = \sum_{i=1}^{n} \frac{p_i^2}{2m_i},$$

and $H = T + U$ is the total energy. Thus the dimension of the phase space of (5) equals $2n$.

Problem 1. State and prove the theorems on existence, uniqueness, continuous and differentiable dependence on initial conditions, and also the extension theorem, for the system (4).

9.5. Remark. The equation of variations. The theorem on differentiability with respect to parameters is not only of theoretical interest, but is also a powerful computational tool.† For example, suppose we can solve a system of differential equations for a certain value of the parameter. Then we can find approximate solutions for neighboring values of the parameter. To do this, we need only calculate the derivative of the solution with respect to the parameter (for the fixed value of the parameter for which we can solve the system). Then it is easy to see that this derivative, regarded as a function of time, is itself a solution of a certain differential equation, called the *equation of variations*. The equation of variations can often be solved without solving the original equation, since it is a (nonhomogeneous) *linear* equation. The effect of all kinds of small perturbations is investigated in various branches of science in this way (by invoking the "method of small parameters").

For example, consider the equation

$$\dot{\mathbf{x}} = \mathbf{v}(\mathbf{x}, \varepsilon)$$

involving a small parameter ε, where $\mathbf{v} = \mathbf{v}_0 + \varepsilon\mathbf{v}_1 + O(\varepsilon^2)$, $\varepsilon \to 0$. By the theorem on differentiability with respect to a parameter, the solution with fixed initial condition can be written in the form

$$\mathbf{x}(t) = \mathbf{x}_0(t) + \varepsilon\mathbf{y}(t) + O(\varepsilon^2),$$

where $\mathbf{x}_0$ is the solution of the "unperturbed" equation

$$\dot{\mathbf{x}} = \mathbf{v}(\mathbf{x}, 0)$$

and $\mathbf{y}$ is the derivative of the solution with respect to the parameter ε at $\varepsilon = 0$. Substituting $\mathbf{x}(t)$ into the original differential equation, we get‡

$$\dot{\mathbf{x}}_0 + \varepsilon\dot{\mathbf{y}} = \mathbf{v}_0(\mathbf{x}_0) + \varepsilon\mathbf{v}_1(\mathbf{x}_0) + \varepsilon\frac{\partial\mathbf{v}_0}{\partial\mathbf{x}}\bigg|_{\mathbf{x}_0}\mathbf{y} + O(\varepsilon^2),$$

a relation valid for all small ε. Therefore the derivatives of both sides of the equation with respect to ε at $\varepsilon = 0$ are equal, i.e.,

$$\dot{\mathbf{y}} = A(t)\mathbf{y} + \mathbf{b}(t)$$

---

† The theorem on differentiation with respect to the initial conditions can thus be used to approximate a bundle of solutions with initial conditions near certain "unperturbed" values for which the solution is known.

‡ Since

$$\mathbf{v}_0(\mathbf{x}) = \mathbf{v}_0(\mathbf{x}_0) + \varepsilon\frac{\partial\mathbf{v}_0}{\partial\mathbf{x}}\bigg|_{\mathbf{x}_0}\mathbf{y} + O(\varepsilon^2)$$

for small ε.

where

$$A(t) = \frac{\partial \mathbf{v}_0}{\partial \mathbf{x}}\bigg|_{\mathbf{x}_0(t)}, \qquad \mathbf{b}(t) = \mathbf{v}_1(\mathbf{x}_0(t)).$$

This is the desired equation of variations. Note that $\mathbf{y}$ also satisfies the initial condition $\mathbf{y}(0) = 0$, since the initial condition for $\mathbf{x}$ is the same for all ε.

In solving problems it is easier to derive the equation of variations as needed, rather than attempt to memorize it.

Problem 1. A body falls vertically in a medium with small resistance depending on both position and velocity:

$$\ddot{x} = -g + \varepsilon F(x, \dot{x}), \qquad \varepsilon \ll 1.$$

Calculate the effect of the resistance on the motion.

Solution. In the absence of resistance ($\varepsilon = 0$), the solution is known:

$$x_0(t) = x(0) + vt - g\frac{t^2}{2}.$$

According to the theorem on differentiation with respect to a parameter, the solution can be written in the form

$$x = x_0 + \varepsilon y(t) + O(\varepsilon^2)$$

for small ε, where y is the derivative of the solution with respect to the parameter ε for $\varepsilon = 0$. Substituting this expression into the original differential equation, we get an equation for y. In fact,

$$\ddot{x}_0 + \varepsilon \ddot{y} = -g + \varepsilon F(x_0, \dot{x}_0) + O(\varepsilon^2), \qquad \varepsilon \to 0,$$

and since this relation holds for all small ε, the coefficient of any power of ε is the same in both sides of the equation. In particular, this gives the following easily solved equation of variations:

$$\ddot{y} = F(x_0(t), \dot{x}_0(t)), \qquad y(0) = \dot{y}(0) = 0.$$

Ans. $x(t) = x_0(t) + \varepsilon \int_0^t \int_0^\tau F(x_0(\xi), \dot{x}_0(\xi))\, d\xi\, d\tau + O(\varepsilon^2).$

Warning. Strictly speaking, our argument is valid only for sufficiently small t, but in fact it is easily justified for any *finite* time interval $|t| \leqslant T$, *provided that ε does not exceed some quantity depending on T (the constant implicit in the term written as $O(\varepsilon^2)$ increases with T).* It is extremely risky to extend the results obtained in this way to an *infinite* time interval: One cannot interchange the limits as $t \to \infty$ and $\varepsilon \to 0$.

Example 1. Consider a bucket of water whose bottom has a small hole of radius ε (Fig. 70). Given any T, there exists a value of ε so small that the bucket remains almost full during the time $t < T$. However, for every fixed $\varepsilon \to 0$, the bucket becomes empty as the time approaches infinity.

Problem 2. As is well known, a body of mass m moving relative to the earth with velocity $\mathbf{v}$

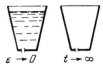

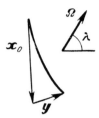

Fig. 70   The asymptotic behavior of the perturbed equation as $\varepsilon \to 0$ and as $t \to \infty$.

Fig. 71   Deflection of a falling body from the vertical.

is subject to a Coriolis force $\mathbf{F} = 2m\mathbf{v} \times \mathbf{\Omega}$, where $\mathbf{\Omega}$ is the angular velocity vector of the earth. A stone is dropped (without initial velocity) into a mine of depth 250 m at the latitude of Leningrad ($\lambda = 60°$). How far from the vertical is the stone deflected by the Coriolis force (Fig. 71)?

Solution. Here we are dealing with the differential equation

$$\ddot{\mathbf{x}} = \mathbf{g} + 2\dot{\mathbf{x}} \times \mathbf{\Omega},$$

depending on the earth's angular velocity $\Omega = 7.3 \times 10^{-5} \text{ sec}^{-1}$ as a parameter. It can be predicted in advance that the Coriolis force is small compared to the weight, and hence Ω can be regarded as a *small* parameter. According to the differentiability theorem, we have

$$\mathbf{x} = \mathbf{x}_0 + \Omega \mathbf{y} + O(\Omega^2)$$

for small Ω, where

$$\mathbf{x}_0 = \mathbf{x}(0) + \mathbf{g}\frac{t^2}{2}.$$

Substituting this expression for $\mathbf{x}$ into the differential equation, we get the equation of variations

$$\ddot{\mathbf{y}} = 2t\mathbf{g} \times \mathbf{\Omega}, \qquad \mathbf{y}_0(0) = \dot{\mathbf{y}}_0(0) = 0,$$

and hence

$$\mathbf{y} = \mathbf{g} \times \mathbf{\Omega}\frac{t^3}{3} = \frac{2t}{3}\mathbf{h} \times \mathbf{\Omega}, \qquad \mathbf{h} = \mathbf{g}\frac{t^2}{2}.$$

Therefore the stone is deflected to the east by

$$\frac{2t}{3}|\mathbf{h}|\,|\mathbf{\Omega}|\cos\lambda \approx \frac{2 \cdot 7}{3} \cdot 250 \cdot 7 \cdot 10^{-5} \cdot \frac{1}{2}\text{m} \approx 4 \text{ cm}.$$

Other examples of the application of the theorems on differentiability with respect to parameters and initial conditions are given in Secs. 12.10 and 26.7.

9.6. Remarks on terminology. Equations of the form (1) and systems of the form (4) are sometimes said to be *in normal form*, or to be *solved with respect to the highest derivatives*. Since these are the only kind of equations and systems considered in this book, the term *system of differential equations* always denotes a system in normal form or a system equivalent to a system in normal form (like the system (5) of Newton's equations).

We also note that the function appearing in the right-hand side of the system (4) can be specified in various ways, e.g., explicitly, implicitly, parametrically, etc.

Example 1. The formula

$$\dot{x}^2 - x = 0$$

is shorthand for *two different* differential equations $\dot{x} = \sqrt{x}$ and $\dot{x} = -\sqrt{x}$, each with the half-line $x > 0$ as its phase space. These equations are determined by two different vector fields, both differentiable for $x > 0$ (Fig. 72).

When an equation is given implicitly, the right-hand side must be treated carefully, with a view to determining its domain of definition and avoiding ambiguous notation.

Example 2. Let $x_1 = r \cos \varphi, x_2 = r \sin \varphi$. Then the formulas $\dot{x}_1 = r, \dot{x}_2 = r\varphi$ do not specify *any* system of differential equations in the plane (x_1, x_2). The same formulas, regarded in any domain of the plane (x_1, x_2) which does not contain the origin of coordinates, lead to *infinitely many* systems of differential equations, corresponding to the infinitely many "branches" of the multiple-valued function φ.

Example 3. By a *Clairaut equation* is meant a differential equation of the form

$$x = \dot{x}t + f(\dot{x}).$$

The Clairaut equation

$$x = \dot{x}t - \frac{\dot{x}^2}{2} \tag{6}$$

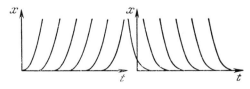

Fig. 72   Integral curves of the two differential equations comprised in the single formula $\dot{x}^2 = x$.

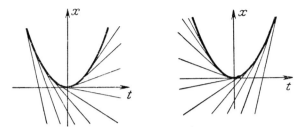

Fig. 73   Integral curves of the two equations written together as the Clairaut equation (6).

is shorthand for *two* different differential equations defined in the domain
$x \leqslant t^2/2$, each satisfying the conditions of the existence and uniqueness
theorems in the domain $x < t^2/2$ under the parabola (Fig. 73). There are
two tangents to the parabola through each point of this domain, and each
tangent consists of two tangent rays. Each of the tangent rays is an integral
curve of one of the two equations given by formula (6).

Problem 1. Investigate the Clairaut equation $x = \dot{x}t - \dot{x}^3$.

10. Phase Curves of Autonomous Systems

We now return to the autonomous case and consider some properties of
solutions of autonomous systems and the corresponding phase curves,
beginning with the following example.

10.1. Time shifts. Consider the equation

$$x^{(n)} = F(x, \dot{x}, \ddot{x}, \ldots, x^{(n-1)}), \tag{1}$$

where F is a differentiable function on the phase space $\mathbf{R}^n$.

Problem 1. Suppose $x = \sin t$ is a solution of equation (1). Prove that
$x = \cos t$ is also a solution.

   This is an immediate consequence of the following

THEOREM. *Let* $\boldsymbol{\varphi} \colon \mathbf{R} \to U$ *be the solution of the autonomous differential equation*

$$\frac{d\mathbf{x}}{dt} = \mathbf{v}(\mathbf{x}) \tag{2}$$

determined by a vector field $\mathbf{v}$ *in the phase space* U, *and let* $h^s \colon \mathbf{R} \to \mathbf{R}$ *be the shift by s
carrying the point* $t \in \mathbf{R}$ *into the point* $t + s \in \mathbf{R}$. *Then* $\boldsymbol{\varphi} \circ h^s \colon \mathbf{R} \to U$ *is a solution
of* (2) *for arbitrary s. In other words, if* $\mathbf{x} = \boldsymbol{\varphi}(t)$ *is a solution of* (2), *then so is*
$\mathbf{x} = \boldsymbol{\varphi}(t + s)$.

Proof. An obvious consequence of the fact that

$$\frac{d\varphi(t+s)}{dt}\bigg|_{t=t_0} = \frac{d\varphi(t)}{dt}\bigg|_{t=t_0+s} = \mathbf{v}(\varphi(t_0+s)) = \mathbf{v}(\varphi(t+s))|_{t=t_0}$$

for arbitrary $t_0 \in \mathbf{R}$, $s \in \mathbf{R}$. ∎

Remark. The theorem immediately implies the analogous assertion for autonomous *systems* and in particular for equation (1). For $s = \pi/2$ we get the solution of the problem posed above.

COROLLARY. *There is one and only one phase curve of the autonomous equation* (2) *going through each point of phase space.*†

Proof. Let $\varphi_1 : \mathbf{R} \to U$, $\varphi_2 : \mathbf{R} \to U$ be two solutions and let $\varphi(t_1) = \varphi(t_2) = \mathbf{x}$. Then the solutions φ_2 and $\varphi_3 = \varphi_1 \circ h^{t_1 - t_2}$ satisfy the same initial condition $\varphi_2(t_2) = \varphi_3(t_2) = \mathbf{x}$, and hence coincide by the uniqueness theorem: $\varphi_2 = \varphi_1 \circ h^{t_1 - t_2}$. But the mappings φ and $\varphi \circ h^s : \mathbf{R} \to U$ have the same image, since the mapping $h^s : \mathbf{R} \to \mathbf{R}$ is one-to-one. Therefore $\varphi_1(\mathbf{R}) = \varphi_2(\mathbf{R})$. ∎

Remark. The phase curves of a nonautonomous equation can intersect without coinciding. Therefore the solutions of nonautonomous equations are best followed along integral curves.

Problem 2. Suppose one and only one phase curve goes through each point of the phase space of the equation $\dot{\mathbf{x}} = \mathbf{v}(t, \mathbf{x})$. Does this imply that the equation is autonomous, i.e., that $\mathbf{v}(t, \mathbf{x})$ is independent of time?
Ans. No.

10.2. Closed phase curves. We already know that distinct phase curves of the autonomous equation (2) do not intersect. We now examine whether a single phase curve can intersect itself.

Let $\varphi_0 : I \to U$ (Fig. 74) be a solution of equation (2) taking the same value $\varphi_0(t_1) = \varphi_0(t_2)$ at two points $t_1 < t_2 \in I$.

THEOREM. *A solution φ_0 such that $\varphi_0(t_1) = \varphi_0(t_2)$ can be extended onto the whole t-axis, and the resulting solution $\varphi : \mathbf{R} \to U$ will have the period $T = t_2 - t_1$, i.e., $\varphi(t + T) = \varphi(t)$ for all t.*

Proof. Every $t \in \mathbf{R}$ can be uniquely represented in the form $t = nT + \tau$, $0 \leqslant \tau < T$. Let $\varphi(t) = \varphi_0(t_1 + \tau)$. Then φ is obviously a periodic func-

---

† Here we have in mind maximal phase curves. By a *maximal phase curve* is meant the image of the mapping $\varphi : I \to U$ where φ is a solution which cannot be extended onto any larger interval I containing I (for example, because I is the whole line (so that the solution is already extended indefinitely) or because $\varphi(t)$ approaches the boundary of the domain U as t approaches the boundary of the interval I).

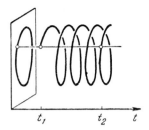

t_1 t_2 t

Fig. 74   A closed phase curve and the corresponding integral curve.

tion with period T. To see that φ is a solution, we note that φ coincides in a neighborhood of every point $t \in \mathbf{R}$ with a translate of the solution φ_0 (this is obvious for points where $\tau > 0$ and follows from the fact that $\varphi_0(t_1) = \varphi_0(t_2)$ for points where $\tau = 0$). Hence φ is a solution, by Theorem 10.1, and the proof is complete, since $\varphi(t_1) = \varphi_0(t_1)$.   ∎

We now consider the set of all periods of the resulting continuous function φ.

LEMMA 1. *The set of all periods of the continuous function* $\varphi : \mathbf{R} \to U$ *is a closed subgroup of the group of real numbers* $\mathbf{R}$.

Proof. If $\varphi(t + T_1) \equiv \varphi(t)$ and $\varphi(t + T_2) \equiv \varphi(t)$, then $\varphi(t + T_1 \pm T_2) \equiv \varphi(t + T_1) \equiv \varphi(t)$, where $\equiv$ indicates identical equality in t.   Moreover, if $T_i \to T$, then

$$\varphi(t + T) \equiv \lim_{i \to \infty} \varphi(t + T_i) \equiv \lim_{i \to \infty} \varphi(t) \equiv \varphi(t),$$

by the continuity of φ.   ∎

LEMMA 2. *Every closed subgroup* G *of the group of real numbers* $\mathbf{R}$ *is either* $\mathbf{R}$ *or* $\{0\}$ *or a set* $\{kT_0, k \in \mathbf{Z}\}$ *of all integral multiples of some number* $T_0 \in \mathbf{R}$.

Proof. If $G \neq \{0\}$, then there exist positive elements in G (if $t < 0$, then $-t > 0$). Let

$$T_0 = \inf \{t : t \in G, t > 0\}.$$

Obviously $0 \leqslant T_0 < \infty$. Suppose $T_0 > 0$. Then T_0 belongs to G, since G is closed. The integral multiples of T_0 also belong to G, since G is a subgroup. Moreover, G contains no other points. In fact, the points kT_0 divide the line $\mathbf{R}$ into intervals $kT_0 < t < (k + 1)T_0$ (Fig. 75). If the group G had an extra element, the element would fall in an interval of the indicated type and then G would contain an element $t - kT_0$ such that $0 < t - kT_0 < T_0$, con-

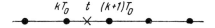

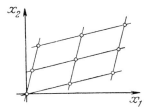

Fig. 75   A closed subgroup of the line.

Fig. 76   A closed subgroup of the plane.

Fig. 77   A limit cycle.

trary to the definition of T_0 as a least upper bound. Hence $T_0 > 0$ implies $G = \{kT_0 : k \in \mathbf{Z}\}$.

We must still consider the case $T_0 = 0$. In this case, given any $\varepsilon > 0$, G contains an element t, $0 < t < \varepsilon$ and hence all points kt, $k \in \mathbf{Z}$. The points kt divide $\mathbf{R}$ into intervals of length less than ε, and hence there are points of G in every neighborhood of any point of $\mathbf{R}$. But then $G = \mathbf{R}$, since G is a closed set. ∎

*Problem 1. Find all closed subgroups of the plane $\mathbf{R}^2$ (Fig. 76), the space $\mathbf{R}^n$, and the group of the circle

$$S^1 = \{z \in \mathbf{C} : |z| = 1\}.$$

Returning to periodic functions, we see that *the set of periods either makes up* the *whole line* (in which case the function is constant) *or else consists of all integral multiples of the smallest period* T_0. Thus a self-intersecting phase curve is either a stationary point or a closed curve which first becomes closed at time T_0, as in the case of a limit cycle (Fig. 77).

Problem 2. Prove that a closed phase curve is diffeomorphic to a circle,[†] unless the curve reduces to a point.

† The definition of a diffeomorphic mapping of one curve onto another is given, for example, in Sec. 33.6.

Hint. The diffeomorphism can be described by the formula

$$\varphi(t) \sim \left(\cos \frac{2\pi t}{T_0}, \sin \frac{2\pi t}{T_0}\right).$$

Nonclosed phase curves can wind around each other in a complicated way, although they cannot intersect each other.

Problem 3. Find the closures of the phase curves of the "double pendulum":

$$\ddot{x}_1 = -x_1, \qquad \ddot{x}_2 = -2x_2.$$

Ans. A point, circles, and tori $S^1 \times S^1$ (see Secs. 24 and 25.6).

Problem 4. Let $\varphi: \mathbf{R} \to U$ be a solution of (2) corresponding to a nonclosed phase curve, so that $\varphi(t_1) \neq \varphi(t_2)$ if $t_1 \neq t_2$. Then the mapping φ of the line R onto the phase curve $\Gamma = \varphi(R)$ is one-to-one, with inverse $\varphi^{-1}: \Gamma \to \mathbf{R}$.
   Is φ^{-1} necessarily continuous?

Hint. See the preceding problem. It can happen that

$$\lim_{i \to \infty} \varphi(t_i) \in \Gamma, \qquad \lim_{i \to \infty} t_i = \infty.$$

11. The Directional Derivative. First Integrals

Many geometric concepts can be described in two ways, either in the language of *points* in space or with the help of functions defined on the space, a duality often found to be useful in various branches of mathematics. In particular, vector fields can be described not only by using curves, but also in terms of *differentiation of functions*. The basic theorems can then be formulated in terms of *first integrals*.

11.1. The derivative in the direction of a vector. Let U be a domain in Euclidean space, x a point of U, and $\mathbf{v}$ a tangent vector, $\mathbf{v} \in TU_x$ (Fig. 78). Let $f: U \to R$ be a differentiable function, and let $\varphi: I \to U$ be any curve leaving x with velocity $\mathbf{v}$, $\varphi(0) = x$. Then the interval I is mapped into the real axis by the composite function

$$f \circ \varphi: I \to \mathbf{R}, \qquad (f \circ \varphi)(t) = f(\varphi(t)),$$

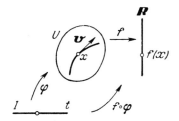

Fig. 78   Derivative of the function f in the direction of the vector $\mathbf{v}$.

which is a real function of a real variable.

Definition. By the *derivative of the function f in the direction of the vector* **v** *is meant the number*

$$L_{\mathbf{v}}f|_x = \frac{d}{dt}\bigg|_{t=0} f \circ \varphi.$$

To justify this definition, we must verify that the number in question does not depend on the choice of the curve φ, but only on the velocity vector **v**. This follows, for example, from the expression for the derivative in terms of the coordinates. By the rule for differentiation of a composite function, we have

$$L_{\mathbf{v}}f|_x = \frac{d}{dt}\bigg|_{t=0} f \circ \varphi = \sum_{i=1}^{n} \frac{\partial f}{\partial x_i}\bigg|_x v_i, \tag{1}$$

where $x_i \colon U \to \mathbf{R}$ is a system of coordinates in the domain U and the v_i are the components of the vector **v** in this system (which are independent of the choice of φ).

11.2. The directional derivative of a vector field. Now let **v** be a vector field in a domain U, so that there is a tangent vector $\mathbf{v}(x) \in TU_x$ at every point $x \in U$. If $f \colon U \to \mathbf{R}$ is a differentiable function, we can form its derivative in the direction of $\mathbf{v}(x)$. This gives a number $L_{\mathbf{v}}f|_x$ at every point of U.

Definition. By the *derivative of the function $f \colon U \to \mathbf{R}$ in the direction of the field* **v** *is meant the new function $L_{\mathbf{v}}f \colon U \to R$ whose value at x equals the derivative of f in the direction of $\mathbf{v}(x)$.*

Example 1. Let $\mathbf{e}_1$ be parallel to the first basis vector of the standard basis of Euclidean space, i.e., the vector with components $1, 0, \ldots, 0$ in a system of coordinates $x_1, x_2, \ldots, x_n$ in U. Then clearly

$$L_{\mathbf{e}_1}f = \frac{\partial f}{\partial x_1}.$$

It follows from (1) that if the function f and the field **v** are of class C^r, then the function $L_{\mathbf{v}}f$ is of class C^{r-1}.

11.3. Properties of the directional derivative. Let F denote the set of all infinitely differentiable functions $f \colon U \to \mathbf{R}$. This set has the natural structure of a real linear space (since addition of functions preserves differentiability) and even of a ring (since a product of differentiable functions is differentiable). Let **v** be an infinitely differentiable vector field. Then the derivative $L_{\mathbf{v}}f$ of the function $f \in F$ in the direction of **v** is again an element of F (the infinite differentiability is essential here!). Thus differentiation in the direction of the field $\mathbf{v}$ is a mapping $L_{\mathbf{v}}: F \to F$ of the ring of infinitely differentiable functions into itself.

Problem 1. Prove the following properties of the operator $L_{\mathbf{v}}$ (except for one of the properties which is false):

1) $L_{\mathbf{v}}(f + g) = L_{\mathbf{v}}f + L_{\mathbf{v}}g$;
2) $L_{\mathbf{v}}(fg) = fL_{\mathbf{v}}g + gL_{\mathbf{v}}f$;
3) $L_{\mathbf{u}+\mathbf{v}} = L_{\mathbf{u}} + L_{\mathbf{v}}$;
4) $L_{f\mathbf{u}} = fL_{\mathbf{u}}$;
5) $L_{\mathbf{u}}L_{\mathbf{v}} = L_{\mathbf{v}}L_{\mathbf{u}}$.

(Here f, g are sufficiently smooth functions, and $\mathbf{u}, \mathbf{v}$ are sufficiently smooth vector fields.)

11.4. Remarks on terminology.

Algebraists apply the term *differentiation* to any mapping of an arbitrary (commutative) ring F into itself which satisfies properties 1) and 2) of the mapping $L_{\mathbf{v}}$. The set of all differentiations of a ring forms a module over the ring.

Thus the vector fields in U form a module over the ring F of infinitely differentiable functions defined in U. Properties 3) and 4) mean that the operation L carrying the vector field $\mathbf{v}$ into the differentiation $L_{\mathbf{v}}$ is a homomorphism of F-modules. Property 5) means that the differentiations $L_{\mathbf{u}}$ and $L_{\mathbf{v}}$ commute (and in general they do not).

Problem 2. Is the homomorphism L an isomorphism?

Analysts call the mapping $L_{\mathbf{v}}: F \to F$ a *linear homogeneous differential operator of the first order*. This designation is explained by the fact that properties 1) and 2) imply that the mapping $L_{\mathbf{v}}: F \to F$ is an **R**-linear operator. In the local coordinates $x_1, \ldots, x_n$ this operator takes the form

$$L_{\mathbf{v}} = v_1 \frac{\partial}{\partial x_1} + \cdots + v_n \frac{\partial}{\partial x_n}$$

(see formula (1)).

11.5. Lie algebras of vector fields.

Problem 3. Prove that the differential operator $L_{\mathbf{a}}L_{\mathbf{b}} - L_{\mathbf{b}}L_{\mathbf{a}}$ is not of the second order (as it appears to be at first glance), but rather of the first order, i.e.,

$$L_{\mathbf{a}}L_{\mathbf{b}} - L_{\mathbf{b}}L_{\mathbf{a}} = L_{\mathbf{c}},$$

where $\mathbf{c}$ is a vector field depending on the fields $\mathbf{a}$ and $\mathbf{b}$.

Comment. The field **c**, denoted by [**a**, **b**], is called the *commutator* or *Poisson bracket* of the fields **a** and **b**.

Problem 4. Prove the following three properties of the commutator:

a) [**a**, **b** + λ**c**] = [**a**, **b**] + λ[**a**, **c**], $\lambda \in \mathbf{R}$ (linearity);

b) [**a**, **b**] + [**b**, **a**] = 0   (antisymmetry);

c) [[**a**, **b**], **c**] + [[**b**, **c**], **a**] + [[**c**, **a**], **b**] = 0   (Jacobi's identity).

Comment. A linear space equipped with a binary operation satisfying the above three conditions is called a *Lie algebra*. Thus vector fields, taken with the operation of commutation, form Lie algebras. Other examples of Lie algebras are the following:

1) Three-dimensional space equipped with the operation of vector multiplication;

2) The space of all $n \times n$ matrices with the operation carrying A, B into $AB - BA$.

Problem 5. Starting from the components of the fields **a** and **b** in some co-ordinate system, find the components of their commutator.

Ans. $[\mathbf{a}, \mathbf{b}]_i = \sum_{j=1}^{n} \left(a_j \dfrac{\partial b_i}{\partial x_j} - b_j \dfrac{\partial a_i}{\partial x_j} \right)$.

*Problem 6. Let g^t be the phase flow determined by the vector field **a** and h^s the flow determined by the field **b**. Prove that the flows commute ($g^t h^s = h^s g^t$) if and only if the commutator of the fields vanishes.

11.6. First integrals. Let **v** be a vector field in a domain U, and let $f: U \to \mathbf{R}$ be a differentiable function.

Definition. The function f is said to be a *first integral*† of the differential equation

$$\dot{x} = \mathbf{v}(x), \qquad x \in U \tag{2}$$

if its derivative in the direction of the vector field **v** vanishes:

$$L_\mathbf{v} f = 0. \tag{3}$$

The following two properties are obviously equivalent to equation (3) and can be taken as the definition of a first integral:

1) *The function f is constant along every solution* $\varphi: I \to U$, i.e., if φ is a solution, then every function $f \circ \varphi: I \to \mathbf{R}$ is a constant;

---

† The strange term *first integral* is a relic of the time when mathematicians still tried to solve all differential equations by integration. In those days, the term integral (or particular integral) was used to designate what we now call a solution.

Fig. 79   A phase curve lies entirely on one level surface of the first integral.

Fig. 80   A system without nonconstant first integrals.

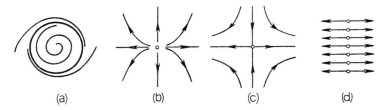

Fig. 81   Which of these systems have nonconstant first integrals?

2) *Every phase curve belongs to one and only one level set† of the function f* (Fig. 79).

Example 1. Consider the following system whose phase space is the whole plane (Fig. 80):

$$\begin{cases} \dot{x}_1 = x_1, \\ \dot{x}_2 = x_2. \end{cases}$$

This system has no first integrals different from a constant. In fact, any first integral is continuous in the whole plane and constant on every ray emanating from the origin, and hence is a constant.

Problem 1. Show that every first integral is constant in a neighborhood of a limit cycle (Fig. 81a) of equation (2).

---

† By the *set of level C* of a function $f: U \to \mathbf{R}$ is meant the full preimage of the point $C \in R$, i.e., the set $f^{-1}C \subset U$.

Problem 2. For what values of k does the system of equations

$$\begin{cases} \dot{x}_1 = x_1, \\ \dot{x}_2 = kx_2, \end{cases} \quad (x_1, x_2) \in \mathbf{R}^2$$

have a nonconstant first integral (Figs. 81b, c, d)?

Nonconstant first integrals are rarely encountered. Hence in those cases where they exist and can be found, they are of great interest.

Example 2. Let H be a differentiable ($r \geqslant 2$ times) function of $2n$ variables $p_1, \ldots, p_n, q_1, \ldots, q_n$. Then by *Hamilton's canonical equations*† we mean the system of $2n$ equations

$$\dot{p}_i = -\frac{\partial H}{\partial q_i}, \qquad \dot{q}_i = \frac{\partial H}{\partial p_i}, \qquad i = 1, \ldots, n. \tag{4}$$

THEOREM (**Law of conservation of energy**). *The function* $H: \mathbf{R}^{2n} \to \mathbf{R}$ *is a first integral of the system of canonical equations* (4).

Proof. It follows from (1) and (4) that

$$L_{\mathbf{v}}H = \sum_{i=1}^{n} \left[\frac{\partial H}{\partial p_i}\left(-\frac{\partial H}{\partial q_i}\right) + \frac{\partial H}{\partial q_i}\frac{\partial H}{\partial p_i} \right] = 0. \quad \blacksquare$$

11.7. Local first integrals. The absence of nonconstant first integrals is related to the topological structure of the collection of phase curves. In general, the phase curves of a system of differential equations do not all stay on the family of level surfaces of any function, and hence there is no non-constant first integral. However, the phase curves do have a simple structure *locally*, in a neighborhood of any nonsingular point, and nonconstant first integrals do exist locally.

Let U be a domain in n-dimensional Euclidean space, let $\mathbf{v}$ be a differentiable vector field in U, and let x be a nonsingular point ($\mathbf{v}(x) \neq 0$).

THEOREM. *There exists a neighborhood V of the point $x \in U$ such that equation* (2) *has $n - 1$ functionally independent‡ first integrals $f_1, \ldots, f_{n-1}$ in V. Moreover, any first integral of (2) in V is a function of $f_1, \ldots, f_{n-1}$.*

Proof. The theorem is obvious for the standard equation

$$\dot{y}_1 = 1, \qquad \dot{y}_2 = \cdots = \dot{y}_n = 0 \tag{5}$$

† It was shown by Hamilton that the differential equations of a great variety of problems encountered in mechanics, optics, calculus of variations, and other branches of science can be written in the form (4).
‡ It will be recalled from calculus that the functions $f_1, \ldots, f_m: U \to \mathbf{R}$ are functionally independent in a neighborhood of a point $x \in U$ if the rank of the derivative $f_*|_x$ of the mapping $f: U \to \mathbf{R}^m$ determined by the functions $f_1, \ldots, f_m$ equals m.

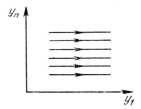

Fig. 82   The coordinate y_n is a first integral.

in $\mathbf{R}^n$ (Fig. 82). In fact, the first integrals are arbitrary differentiable functions of the coordinates $y_2, \ldots, y_n$, and the coordinates $y_2, \ldots, y_n$ give us $n - 1$ functionally independent first integrals. The same is true for equation (5) in any convex domain W of the space $\mathbf{R}^n$.† By the basic theorem (Sec. 7.1), equation (2) is of the form (5) in some neighborhood of the point x in suitable coordinates y, and this neighborhood can be regarded as a convex domain in the coordinates y (otherwise replace it by a smaller convex neighborhood). It remains only to note that the property of a function being a first integral and the property of functional independence are both independent of the coordinate system. ∎

11.8. Time-dependent first integrals. Let $f: \mathbf{R} \times U \to \mathbf{R}$ be a differentiable function in the extended phase space of the equation

$$\dot{x} = \mathbf{v}(t, x), \qquad t \in \mathbf{R}, \qquad x \in U, \tag{6}$$

which is in general nonautonomous (the right-hand side $\mathbf{v}(t, x)$ is assumed to be differentiable). Then the function f is said to be a *time-dependent first integral* if it is a first integral of the autonomous system obtained from (6) by adjoining the equation $\dot{t} = 1$:

$$\dot{X} = \mathbf{V}(X), \qquad X \in \mathbf{R} \times U, \qquad X = (t, x), \qquad \mathbf{V}(t, x) = (1, \mathbf{v}).$$

In other words, *every integral curve of equation* (6) *lies entirely on one level set of the function* f (Fig. 83).

The vector field V does not vanish. It follows from the preceding theorem that *equation* (6) *has n functionally independent* (*time-dependent*) *first integrals* $f_1, \ldots, f_n$ *in some neighborhood of every point* (t, x) *and that every* (*time-dependent*) *first integral of* (6) *can be expressed in terms of* $f_1, \ldots, f_n$ *in this neighborhood.*

In particular, the autonomous equation (2) with an n-dimensional phase

† A domain in $\mathbf{R}^n$ is said to be *convex* if whenever it contains two points, it also contains the line segment joining the two points. Give an example of a first integral of (5) which does not reduce to a function of $y_2, \ldots, y_n$ in a nonconvex domain W of the space $\mathbf{R}^n$.

Fig. 83   Integral curves on a level surface of a time-dependent first integral.

space has n time-dependent functionally independent first integrals in the neighborhood of *any* (not necessarily nonsingular) point.

Problem 1. Suppose every solution of equation (6) can be extended onto the whole t-axis. Prove that equation (6) then has n functionally independent (time-dependent) first integrals in the whole extended phase space, in terms of which we can express every (time-dependent) first integral.

By a first integral of a differential equation (or of a system of differential equations) of arbitrary order is meant a first integral of the equivalent system of first-order equations.

12. Conservative Systems with One Degree of Freedom

As an example of the application of first integrals to the investigation of differential equations, we now consider a frictionless mechanical system with one degree of freedom.

12.1. Definitions. By a *conservative system with one degree of freedom* is meant a system described by the differential equation

$$\ddot{x} = F(x), \tag{1}$$

where F is a differentiable function defined on an interval I of the real x-axis. Equation (1) is equivalent to the system

$$\begin{cases} \dot{x}_1 = x_2, \\ \dot{x}_2 = F(x_1), \end{cases} \qquad (x_1, x_2) \in I \times \mathbf{R}. \tag{2}$$

The following terminology is customary in mechanics:

I the configuration space;

$x_1 = x$ the coordinate;

$x_2 = \dot{x}$ the velocity;

$\ddot{x}$ the acceleration;

$I \times \mathbf{R}$ the phase space;

(1)   Newton's equation;

F the force field;

$F(x)$ the force.

We also consider the following functions defined on phase space:

$T = \frac{1}{2}\dot{x}^2 = \frac{1}{2}x_2^2$ the *kinetic energy*;

$U = -\int_{x_0}^{x} F(\xi)\,d\xi$ the *potential energy*;

$E = T + U$ the *total mechanical energy*.

Obviously $F(x) = -\partial U/\partial x$, so that *the potential energy determines the system.*

Example 1. For the pendulum (Sec. 1.6), we have

$$\ddot{x} = -\sin x,$$

where x is the angle of deviation, so that

$$F(x) = -\sin x, \qquad U(x) = -\cos x$$

(Fig. 84). Moreover

$$\ddot{x} = -x, \qquad F(x) = -x, \qquad U(x) = \frac{1}{2}x^2,$$

for small oscillations of the pendulum, while

$$\ddot{x} = x, \qquad F(x) = x, \qquad U(x) = -\frac{1}{2}x^2$$

for small oscillations of the inverted pendulum (Fig. 85).

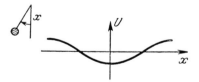

Fig. 84   Potential energy of the pendulum.

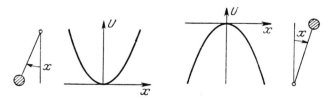

Fig. 85   Potential energy of the pendulum near the lower and upper equilibrium positions.

12.2. The law of conservation of energy.

THEOREM. *The total energy E is a first integral of the system* (2).

Proof. We need only note that

$$\frac{d}{dt}\left[\frac{1}{2}x_2^2(t) + U(x_1(t))\right] = x_2\dot{x}_2 + U'\dot{x}_1 = x_2 F(x_1) - F(x_1)x_2 = 0. \quad \blacksquare$$

With the help of this theorem, an equation of the form (1), for example the pendulum equation, can be investigated and explicitly solved "in quadratures."

12.3. Level curves of the energy. Turning to the phase curves of the system (2), we note that each such curve lies entirely on one level set of the energy. We now study these level sets.

THEOREM. *The level set of the energy*

$$\{(x_1, x_2): \tfrac{1}{2}x_2^2 + U(x_1) = E\}$$

is a smooth curve in a neighborhood of each of its points, with the exception of the equilibrium positions, i.e., the points (x_1, x_2) where

$$F(x_1) = 0, \qquad x_2 = 0.$$

Proof. We use the implicit function theorem, observing that

$$\frac{\partial E}{\partial x_1} = -F(x_1), \qquad \frac{\partial E}{\partial x_2} = x_2.$$

If one of these derivatives is nonvanishing, then the set of level E is the graph of a differentiable function of the form $x_1 = x_1(x_2)$ or $x_2 = x_2(x_1)$ in a neighborhood of the point in question. $\blacksquare$

Note that the exceptional points (x_1, x_2) figuring in the theorem, where $F(x_1) = 0$ and $x_2 = 0$, are just the stationary points (equilibrium positions) of the system (2) as well as the singular points of the vector field of the phase velocity. Moreover, the same points are the critical points of the total energy $E(x_1, x_2)$, while the points where $F(x_1) = 0$ are the critical points[†] of the potential energy U.

To draw the level curves of the energy, it is useful to think of a bead sliding in a "potential well" U (Fig. 86).

Suppose the total energy has a fixed value E. Since the potential energy cannot exceed the total energy, the projection onto configuration space

---

† By a *critical point* of a function is meant a point at which the total differential of the function vanishes. The value of the function at such a point is called a *critical value*.

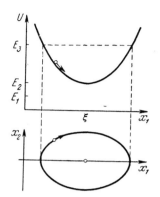

Fig. 86   A bead in a potential well and the corresponding phase curve.

(the x_1-axis) of the level curve of energy E lies in the set $\{x_1 \in I \colon U(x_1) \leqslant E\}$ of points at which the value of the potential energy does not exceed E (the bead cannot go higher than the level E in the potential well). Moreover, the larger the velocity (in absolute value), the smaller the potential energy, since $|x_2| = \sqrt{2(E - U(x_1))}$, i.e., the bead picks up velocity as it falls into the well and loses velocity as it rises from the bottom of the well. Note that the velocity vanishes at the "turning points," where $U(x_1) = E$.

It follows from the evenness of the energy with respect to x_2 that the level curve of the energy is symmetric with respect to the x_1-axis (the bead traverses each point twice in opposite directions with the same speed).

These simple considerations suffice to allow us to sketch level curves of the energy for systems with various potentials U. First we consider the simplest case (an infinitely deep potential well with one attractive center ξ), where $F(x)$ decreases monotonically: $F(\xi) = 0$, $I = \mathbf{R}$ (Fig. 86).

If the value E_1 of the total energy is smaller than the minimum E_2 of the potential energy, the set of level $E = E_1$ is empty (the motion of the bead is physically impossible). The set of level $E = E_2$ then consists of the single point $(\xi, 0)$ (the bead rests at the bottom of the well).

If the value E_3 of the total energy is larger than the critical value $E_2 = U(\xi)$, the set of level $E = E_3$ is a symmetric smooth closed curve surrounding the equilibrium position $(\xi, 0)$ in the phase plane (the bead slides backward and forward in the well, rising to the height E_3, at which time the velocity vanishes, then falling back into the well and going through $(\xi, 0)$, at which time the velocity is maximum, afterwards rising again on the other side, and so on).

To study more complicated cases, we proceed in the same way, i.e., we progressively increase the values of the total energy E, stopping at the values

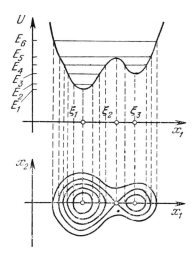

Fig. 87   Level curves of the energy for a potential with two wells.

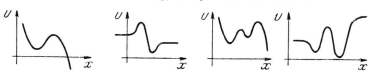

Fig. 88   What is the appearance of the level curves of the energy for each of these potentials?

of E equal to the critical values of the potential energy U (where $U'(\xi) = 0$) and in each case examining the curves with values of E a little smaller and a little larger than the critical value.

Example 1. Suppose the potential energy U has three critical points, a minimum ξ_1, a local maximum ξ_2, and a local minimum ξ_3. Then Fig. 87 shows the level curves corresponding to the values $E_1 = U(\xi_1)$, $U(\xi_1) < E_2 < U(\xi_3)$, $E_3 = U(\xi_3)$, $U(\xi_3) < E_4 < U(\xi_2)$, $E_5 = U(\xi_2)$, $E_6 > U(\xi_2)$.

Problem 1. Sketch level curves of the energy for the pendulum equation $\ddot{x} = -\sin x$ and for the pendulum equations near the lower and upper equilibrium positions ($\ddot{x} = -x$ and $\ddot{x} = x$).

Problem 2. Sketch level curves of the energy for the *Kepler potential*†

$$U = -\frac{1}{x} + \frac{C}{x^2}$$

and for the potentials shown in Fig. 88.

† The change in distance between a planet (or comet) and the sun is described by Newton's equation with this potential.

12.4. Level curves of the energy near a singular point. In studying the behavior of level curves near a critical value of the energy, it is useful to keep the following facts in mind:

Remark 1. If the potential energy is a quadratic form $U = \frac{1}{2}kx^2$, then the level curves of the energy are second-order curves $2E = x_2^2 + kx_1^2$.

In the attractive case, we have $k > 0$ and the critical point 0 is a minimum of the potential energy (Fig. 89). The level curves of the energy are then homothetic ellipses centered at 0.

In the repulsive case, we have $k < 0$ and the critical point 0 is a maximum of the potential energy (Fig. 90). The level curves of the energy are then homothetic hyperbolas centered at 0, together with the pair of asymptotes $x_2 = \pm\sqrt{k}x_1$. These asymptotes are also called *separatrices*, since they separate hyperbolas of different types from one another.

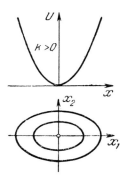

Fig. 89   Level curves of the energy for an attractive quadratic potential.

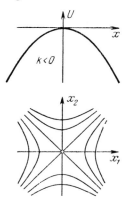

Fig. 90   Level curves of the energy for a repulsive quadratic potential.

Remark 2. The increment of a function $f(x)$ is a quadratic form in a neighborhood of a nondegenerate critical point, provided only that coordinates are suitably chosen.

Here we assume that $f(0) = 0$ and that the derivatives $f'(0)$ and $f''(0)$ exist. The point 0 is a critical point of f if $f'(0) = 0$, and the critical point 0 is then said to be *nondegenerate* if $f''(0) \neq 0$.

LEMMA 1 (**Morse**). *In a neighborhood of a nondegenerate critical point 0, the co-ordinate y can be chosen in such a way that*

$$f = Cy^2, \qquad c = \operatorname{sgn} f''(0).$$

Of course, $y = \operatorname{sgn} x \sqrt{|f(x)|}$ is such a coordinate, and the assertion consists in showing that the correspondence $x \mapsto y$ is diffeomorphic in a neighborhood of 0.

In proving Morse's lemma, we make use of the following proposition:

LEMMA 2 (**Hadamard**).† *Let f be a differentiable function (of class C^r) such that both f and its derivative f' vanish at the point $x = 0$. Then $f(x) = xg(x)$, where g is a differentiable function (of class C^{r-1} in a neighborhood of the point $x = 0$).*

Proof. We need merely note that

$$f(x) = \int_0^1 \frac{df(tx)}{dt}\, dt = \int_0^1 f'(tx)x\, dt = x \int_0^1 f'(tx)\, dt,$$

where

$$g(x) = \int_0^1 f'(tx)\, dt$$

is a function of class C^{r-1}. ∎

Applying Hadamard's lemma twice to the function f figuring in Morse's lemma, we find that $f = x^2 \varphi(x)$, where $2\varphi(0) = f''(0) \neq 0$. Hence $y = x\sqrt{|\varphi(x)|}$ and Morse's lemma is proved, since the function $\sqrt{|\varphi(x)|}$ is differentiable ($r - 2$ times if f is of class C^r) in a neighborhood of the point $x = 0$.

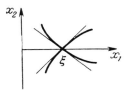

Fig. 91   Tangents to the separatrices of a repulsive singular point.

† Both lemmas can be extended to the case of functions of several variables.

Thus in a neighborhood of a nondegenerate critical point the level curves of the energy become either ellipses or hyperbolas under a diffeomorphic change of the system of coordinates (x_1, x_2).

Problem 1. Find the tangents to the separatrices of a repulsive singular point $(U''(\xi) < 0)$.

Ans. $x_2 = \pm \sqrt{|U''(\xi)|} \, (x_1 - \xi)$ (Fig. 91).

12.5. Extension of solutions of Newton's equations. Suppose the potential energy is defined on the whole x-axis. Then the law of conservation of energy immediately implies the following

THEOREM. *If the potential energy U is positive everywhere,†️ then every solution of the equation*

$$\ddot{x} = -\frac{dU}{dx} \tag{1'}$$

can be extended indefinitely.

Example 1. If $U = -\frac{1}{2}x^4$, the solution $x = 1/(t-1)$ cannot be extended up to $t = 1$.

First we prove the following "a priori estimate":

LEMMA. *If a solution exists for $|t| < \tau$, then it satisfies the inequalities*

$$|\dot{x}(t)| \leqslant \sqrt{2E_0}, \qquad |x(t) - x(0)| < \sqrt{2E_0} \, |t|,$$

where

$$E_0 = \tfrac{1}{2}\dot{x}^2(0) + U(x(0))$$

is the initial value of the energy.

Proof. According to the law of conservation of energy,

$$\tfrac{1}{2}\dot{x}^2(t) + U(x(t)) = E_0,$$

and since $U > 0$, the first inequality is proved. The second inequality follows from the first, since

$$x(t) - x(0) = \int_0^t \dot{x}(\theta) \, d\theta. \quad \blacksquare$$

Proof of the theorem. Let T be an arbitrary positive number, and let Π (Fig. 92) be the rectangle

$$|x_1 - x_1(0)| \leqslant 2\sqrt{2E_0} \, T, \qquad |x_2| \leqslant 2\sqrt{2E_0}$$

---

†️ Naturally, changing the potential energy U by a constant does not change equation (1'). Hence it is only essential that U be bounded from below.

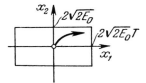

Fig. 92   The rectangle which the phase point cannot leave in time T.

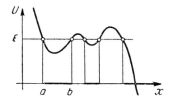

Fig. 93   The set of points x where $U(x) < E$ (E a noncritical energy level).

in the phase plane. Consider the parallelepiped $|t| \leqslant T$, $(x_1, x_2) \in \Pi$ in the extended phase space (x_1, x_2, t). By the extension theorem, the solution can be extended up to the boundary of the parallelepiped. It follows from the lemma that the solution can leave the parallelepiped only through those faces on which $|t| = T$. Hence the solution can be extended up to arbitrary $t = \pm T$, and hence can be extended indefinitely.   ∎

Problem 1. Prove the possibility of indefinitely extending the solutions of the system of Newton's equations

$$m_i \ddot{x}_i = -\frac{\partial U}{\partial x_i}, \qquad i = 1, \ldots, N, \qquad m_i > 0, \qquad x \in \mathbf{R}^N$$

in the case of positive potential energy ($U > 0$).

12.6. Noncritical level curves of the energy. Suppose the potential energy U is defined on the whole x-axis, and let E be a noncritical value of the energy, i.e., let E be different from any of the values of the function U at its critical points. Consider the set of points $\{x : U(x) < E\}$ where the value of U is less than E. Since U is continuous, this set (Fig. 93) consists of a finite or countable number of intervals (two of these intervals may extend to infinity). At the end points of the intervals $U(x) = E$, and hence $U'(x) \neq 0$ since E is a noncritical value. Every point of the set $\{x : U(x) = E\}$ is for this reason the end point of precisely one interval in which $U(x) < E$. Therefore the whole set $\{x : U(x) < E\}$ is either the entire x-axis or the union of no more than countably many pairwise disjoint closed intervals, possibly together with one or two rays extending to infinity. In the following

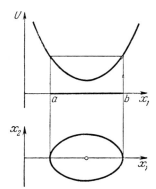

Fig. 94   A phase curve diffeomorphic to a circle.

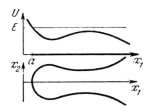

Fig. 95   A phase curve diffeomorphic to a line.

theorem, we consider one of these intervals $a \leqslant x \leqslant b$ (Fig. 94), where $U(a) = U(b) = E$ and $U(x) < E$ for $a < x < b$.

THEOREM. *The equation*

$$\tfrac{1}{2}x_2^2 + U(x_1) = E, \qquad a \leqslant x_1 \leqslant b$$

determines a smooth curve in the plane (x_1, x_2). This curve is diffeomorphic to a circle and is a phase curve of the system (2). *Similarly, the ray $a \leqslant x < \infty$ (or $-\infty < x \leqslant b$), where $U(x) < E$, is the projection onto the x_1-axis of a phase curve diffeomorphic to a straight line (Fig. 95). Finally, in the case where $U(x) < E$ on the whole line, the set of level E consists of two phase curves*

$$x_2 = \pm\sqrt{2(E - U(x_1))}.$$

Thus the set of level E, where the energy E is noncritical, consists of a finite or countable number of smooth phase curves.

12.7. Proof of Theorem 12.6. The law of conservation of energy allows us to solve Newton's equation explicitly. In fact, for a fixed value of the total

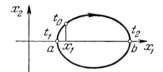

Fig. 96   The phase point traverses half the phase curve (from a to b) in a finite time $T/2 = t_2 - t_1$.

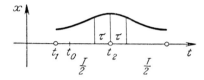

Fig. 97   Use of reflection to extend the solution of Newton's equation.

energy E, the magnitude (but not the sign) of the velocity $\dot{x}$ is determined by the position x, since

$$\dot{x} = \pm\sqrt{2(E - U(x))}, \tag{3}$$

and we already know how to solve this one-dimensional equation.

Let (x_1, x_2) be a point of our level set, where $x_2 > 0$ (Fig. 96). Making use of (3), we look for a solution φ of equation (1) satisfying the initial condition $\varphi(t_0) = x_1, \dot{\varphi}(t_0) = x_2$, obtaining

$$t - t_0 = \int_{x_1}^{\varphi(t)} \frac{d\xi}{\sqrt{2(E - U(\xi))}} \tag{4}$$

for t near t_0. We now observe that the integral

$$\frac{T}{2} = \int_a^b \frac{d\xi}{\sqrt{2(E - U(\xi))}}$$

converges, since $U'(a) \neq 0$, $U'(b) \neq 0$. Therefore (4) defines a continuous function φ on some interval $t_1 \leqslant t \leqslant t_2$ with $\varphi(t_1) = a$, $\varphi(t_2) = b$. This function satisfies Newton's equation everywhere (Fig. 97).

The interval (t_1, t_2) is of length $T/2$. We now extend φ onto the next interval of length $T/2$ by using symmetry considerations: $\varphi(t_2 + \tau) = \varphi(t_2 - \tau), 0 \leqslant \tau \leqslant T/2$, further extending φ by periodicity: $\varphi(t + T) \equiv \varphi(t)$. The resulting function, defined on the whole line, satisfies Newton's equation everywhere, and moreover $\varphi(t_0) = x_1, \dot{\varphi}(t_0) = x_2$. Thus we have constructed a solution of the system (2) satisfying the initial condition (x_1, x_2), which turns out to be periodic with period T. The corresponding

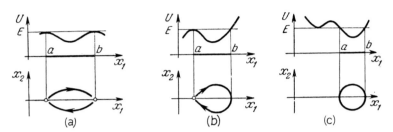

Fig. 98   Decomposition of a critical level curve of the energy into phase curves.

closed phase curve is just the part of the set of level E lying over the interval $a \leqslant x \leqslant b$. This curve is diffeomorphic to a circle, like every closed phase curve (see Sec. 10).

The case where the interval extends to infinity (in one direction or the other) is simpler than the case just considered, and is left as an exercise. ∎

12.8. Critical level curves. The structure of critical level curves can be more complicated. Note that such curves contain fixed points (x_1, x_2) (where $U'(x_1) = 0$, $x_2 = 0$), each of which is itself a phase curve. If $U(x) < E$ everywhere on the interval $a \leqslant x \leqslant b$, except for $U(a) = U(b) = E$, and if both end points are critical points, so that $U'(a) = U'(b) = 0$, then both open arcs

$$x_2 = \pm \sqrt{2(E - U(x_1))}, \qquad a < x_1 < b$$

(Fig. 98a) are phase curves. The time taken by the phase point to traverse such an arc is infinite (Theorem 12.5 + uniqueness).

If $U'(a) = 0$, $U'(b) \neq 0$ (Fig. 98b), the equation

$$\tfrac{1}{2}x_2^2 + U(x_1) = E, \qquad a < x_1 \leqslant b$$

determines a nonclosed phase curve. Finally, if $U'(a) \neq 0$, $U'(b) \neq 0$ (Fig. 98c), then the part of the critical level set lying over the interval $a \leqslant x \leqslant b$ is a closed phase curve, just as in the case of a noncritical level E.

12.9. Example. The above considerations will now be applied to the pendulum equation

$$\ddot{x} = -\sin x,$$

with potential energy $U(x) = -\cos x$ (Fig. 99) and critical points $x_1 = k\pi$, $k = 0, \pm 1, \ldots$. The closed phase curves resemble ellipses near the point $x_1 = 0$, $x_2 = 0$, and these curves correspond to small oscillations of the pendulum. The period T of the oscillations depends only slightly on the amplitude, as long as the amplitude is small. For larger values of the energy

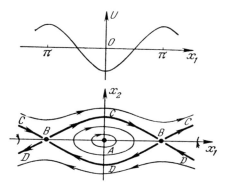

Fig. 99   Phase curves of the pendulum equation $\ddot{x} = -\sin x$.

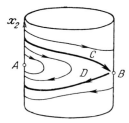

Fig. 100   Cylindrical phase space of the pendulum.

constant, we get larger closed curves, until the energy reaches a critical value equal to the potential energy of the pendulum in the upside-down position. The period of the oscillations then increases (since the time of motion along the separatrices making up the critical level set is infinite).

For larger values of the energy we get nonclosed curves on which x_2 does not change sign, i.e., the pendulum rotates rather than oscillates, achieving the largest value of its velocity at the lower position and the smallest value at the upper position. Note that values of x_1 differing by $2k\pi$ correspond to identical positions of the pendulum. Therefore it is natural to choose the cylinder $(x_1 \bmod 2\pi, x_2)$ rather than the plane (x_1, x_2) as the phase space of the pendulum (Fig. 100).

Taking the picture already drawn in the plane and wrapping it around the cylinder, we get the phase curves of the pendulum on the surface of the cylinder. They are all closed smooth curves, except for two fixed points A, B (the lower and upper equilibrium positions) and two separatrices C, D.

Problem 1. Draw graphs of the functions $x_1(t)$ and $x_2(t)$ for the solution with energy near but somewhat below the critical energy in the upper position.

Fig. 101 The angle of deviation of the pendulum and the velocity of its motion for amplitudes near π.

Ans. See Fig. 101. The functions $x_1(t)$ and $x_2(t)$ can be expressed in terms of sn and cn (the elliptic sine and elliptic cosine). As E approaches the lower critical value, the oscillations of the pendulum become approximately harmonic, with sn and cn going into sin and cos.

Problem 2. At what rate does the period of the oscillations of a pendulum approach infinity, as the energy E approaches the upper critical value E_1?

Ans. At a logarithmic rate $(\sim C \ln (E_1 - E))$.

Hint. See formula (4).

12.10. Small perturbations of a conservative system.

Having investigated the motions of a conservative system, we can now use the theorem on differentiability with respect to a parameter (Sec. 9.5) to study neighboring systems of a general form. In doing so, we encounter a qualitatively new phenomenon of great importance in the applications, i.e., *auto-oscillations* or *self-excited oscillations*.

Problem 1. Investigate the phase curves of the system

$$\begin{cases} \dot{x}_1 = x_2 + \varepsilon f_1(x_1, x_2), \\ \dot{x}_2 = -x_1 + \varepsilon f_2(x_1, x_2), \end{cases} \qquad \varepsilon \ll 1, \quad x_1^2 + x_2^2 \leqslant R^2$$

differing only slightly from the system of equations for small oscillations of a pendulum.

Solution. For $\varepsilon = 0$ we get the equations for small oscillations of a pendulum. By the theorem on differentiability with respect to a parameter, the solution (on a finite time interval) differs by a correction of order ε from the harmonic oscillations

$$x_1 = A \cos (t - t_0), \qquad x_2 = -A \sin (t - t_0),$$

provided ε is small. Hence, for sufficiently small $\varepsilon = \varepsilon(T)$, the phase point stays near the circle of radius A during the interval T.

Unlike the conservative case ($\varepsilon = 0$), the phase curve is not necessarily closed for $\varepsilon \neq 0$, and it may have the form of a spiral (Fig. 102), with a small distance (of order ε) between neighboring turns. To determine whether the phase curve approaches the origin of coordinates or recedes from the origin, we consider the increment of the energy $E = \frac{1}{2}x_1^2 + \frac{1}{2}x_2^2$ after one circuit around the origin. We are particularly interested in the sign of this increment, which is positive on the expanding (unwinding) spiral, negative on the contracting (tightening) spiral, and zero on the limit cycle itself. We now deduce an approximate expression, namely formula (6), for the energy increment.

The derivative of the energy in the direction of our vector field is easily evaluated and is proportional to ε:

$$\dot{E}(x_1, x_2) = \varepsilon(x_1 f_1 + x_2 f_2). \tag{5}$$

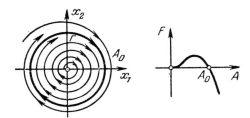

Fig. 102   Phase curves of the van der Pol equation and the increment of energy after one circuit around the origin.

To calculate the energy increment after one circuit, this function should be integrated along a turn of the phase trajectory, but the latter is unfortunately not known. But, as already explained, the turn is close to a circle, and hence, to within an accuracy of $O(\varepsilon^2)$, the integral can be taken along the circle S of radius A:

$$\Delta\varepsilon = \varepsilon\int_0^{2\pi} \dot{E}(A\cos t,\ -A\sin t)\ dt + O(\varepsilon^2).$$

Substituting (5) into this formula, we get[†]

$$\Delta E = \varepsilon F(A) + O(\varepsilon^2) \tag{6}$$

where

$$F(A) = \oint f_1\ dx_2 - f_2\ dx_1$$

(the integral is taken along a circle of radius A traversed in the counterclockwise direction).

Once having calculated the function $F(A)$, we can investigate the behavior of the phase curves. If the function F is positive, the energy increment ΔE after one circuit is also positive (for small positive ε). In this case, the phase curve is an unwinding spiral, and the system executes increasing oscillations. On the other hand, if $F < 0$, then $\Delta E < 0$ and the phase spiral is contracting. In the latter case, the oscillations damp out.

It can happen that the function $F(A)$ changes sign (Fig. 102). Suppose $F(A)$ has a simple zero A_0. Then for small ε the equation

$$\Delta E(x_1, x_2) = 0$$

is satisfied by a closed curve Γ in the phase plane, near the circle of radius A_0 (this follows from the implicit function theorem). Obviously Γ is a closed phase curve, i.e., a limit cycle of our system.

The sign of the derivative

$$F' = \frac{dF}{dA}\bigg|_{A=A_0}$$

determines whether neighboring phase curves wind onto the limit cycle or unwind from it. The cycle is *unstable* if $\varepsilon F' > 0$ and *stable* if $\varepsilon F' < 0$. In fact, in the first case the energy increase after one circuit is greater than zero if the phase curve lies outside the cycle and less than zero if it lies inside the cycle; hence the phase curve always moves away from the

† Here we use the fact that $dx_1 = x_2\ dt$, $dx_2 = -x_1\ dt$ along S.

cycle. However, in the second case, the phase curves approach the cycle both from the inside and from the outside, as in Fig. 102.

Example 1. Consider the equation

$$\ddot{x} = -x + \varepsilon \dot{x}(1 - x^2),$$

called the *van der Pol equation*. Evaluating the integral (6) with $f_1 = 0$, $f_2 = x_2(1 - x_1^2)$, we get

$$F(A) = \pi\left(A^2 - \frac{A^4}{4}\right).$$

This function has a simple zero $A_0 = 2$ (Fig. 102), and is positive for smaller A and negative for larger A. Therefore for small ε the van der Pol equation has a stable limit cycle, close to the circle $x^2 + \dot{x}^2 = 4$ in the phase plane.

Suppose we compare the motion of the original conservative system ($\varepsilon = 0$) with what happens for $\varepsilon \neq 0$. In the conservative system, there can occur oscillations of arbitrary amplitude (all the phase curves are closed), with the amplitude determined only by the initial conditions. In the nonconservative system ($\varepsilon \neq 0$), qualitatively different phenomena are possible, for example, a stable limit cycle. In this case, very different initial conditions lead to the establishment of a periodic oscillation of one and the same completely determined amplitude. The resulting steady-state regime is said to be *auto-oscillatory*.

*Problem 2. Investigate the auto-oscillatory motions of a pendulum with small friction subject to the action of a constant torque M:

$$\ddot{x} + \sin x + \varepsilon \dot{x} = M.$$

Hint. This problem is analyzed in detail for arbitrary ε and M in A. A. Andronov, A. A. Vitt, and S. E. Khaikin, *Theory of Oscillations* (in Russian), Moscow (1959), Chap. 7.

3 Linear Systems

Linear systems are almost the only large class of differential equations for which there exists a definitive theory. This theory is essentially a branch of linear algebra, and allows us to solve all autonomous linear equations.

The theory of linear equations is also useful as a first approximation to the study of nonlinear problems. For example, it allows us to investigate the stability of equilibrium positions and the topological classification of singular points of vector fields in the nondegenerate cases.

13. Linear Problems

We begin by considering two examples of situations in which linear equations arise.

13.1. Example: Linearization. Consider the differential equation determined by a vector field in phase space. We already know that the field has a simple structure in a neighborhood of a nonsingular point ($\mathbf{v} \neq 0$), i.e., it is rectified by a diffeomorphism. We now consider the structure of the field in a neighborhood of a singular point, namely a point where the field vector vanishes. Such a point x_0 is a stationary point of our equation. If the equation describes some physical process, then x_0 is a stationary state of the process, namely its "equilibrium position." Therefore studying a neighborhood of the singular point means studying how the process evolves when its initial conditions deviate slightly from their equilibrium values (consider, for example, the upper and lower equilibrium positions of the pendulum).

To investigate the vector field in a neighborhood of a point x_0 where the field vector vanishes, it is natural to make a Taylor series expansion of the field in the given neighborhood. The first term of the Taylor series is linear, and the process of dropping the remaining terms is called *linearization*. The linearized vector field can be regarded as an example of a vector field with a singular point x_0. On the other hand, it might be expected that the behavior of the linearized equation is close to that of the original equation (since small quantities of higher order are dropped in making the linearization). Of course, the problem of the relation between the solutions of the original equation and those of the linearized equation requires special investigation. This investigation is based on a detailed analysis of the linear equation, a topic which will be our first concern.

Problem 1. Show that *linearization is an invariant operation, i.e., an operation which is independent of the coordinate system.*

More exactly, suppose the field $\mathbf{v}$ in the domain U is given in the system of coordinates x_i by components $v_i(x)$, and let the singular point have coordinates $x_i = 0$, so that $v_i(x) = 0$, $i = 1, \ldots, n$. Then the original equation takes the form of a system

$$\dot{x}_i = v_i(x), \qquad i = 1, \ldots, n.$$

The *linearized equation* is now defined as the equation

$$\dot{\xi}_i = \sum_{j=1}^{n} a_{ij}\xi_j, \qquad i = 1, \ldots, n, \qquad a_{ij} = \left.\frac{\partial v_i}{\partial x_j}\right|_{x=0}.$$

Consider the tangent vector $\boldsymbol{\xi} \in TU_0$ with components ξ_i, $i = 1, \ldots, n$. Then the linear equation can be written in the form

$$\dot{\boldsymbol{\xi}} = A\boldsymbol{\xi},$$

where A is the linear mapping $A: TU_0 \to TU_0$ specified by the matrix (a_{ij}). It is now asserted that *the mapping A does not depend on the system of coordinates x_i figuring in its definition.*

Problem 2. Linearize the pendulum equation $\ddot{x} = -\sin x$ near the equilibrium position $x_0 = k\pi$, $\dot{x}_0 = 0$.

13.2. Example: One-parameter groups of linear transformations of $\mathbf{R}^n$.
Another problem leading at once to linear differential equations is the problem of describing one-parameter groups of linear transformations of the linear space $\mathbf{R}^n$.

First we note that *it is natural to identify the tangent space to the linear space $\mathbf{R}^n$ at any point with the linear space itself.* In fact, we identify the element $\dot{\phi}$ of the tangent space $T\mathbf{R}^n_x$, whose representative is the curve $\varphi: I \to \mathbf{R}^n$, $\varphi(0) = x$, with the vector

$$\mathbf{v} = \lim_{t \to 0} \frac{\varphi(t) - x}{t} \in \mathbf{R}^n$$

of the space $\mathbf{R}^n$ itself (the correspondence $\mathbf{v} \to \dot{\phi}$ is one-to-one).

This identification depends on the structure of the linear space $\mathbf{R}^n$ and is *not* preserved under diffeomorphisms. However, in the linear problems which will now concern us (for example, in the problem of one-parameter groups of linear transformations), the structure of the linear space in $\mathbf{R}^n$ is fixed once and for all. Therefore *we now make the identification $T\mathbf{R}^n_x \equiv \mathbf{R}^n$ until such time as we return to nonlinear problems.*

Let $\{g^t, t \in \mathbf{R}\}$ be a one-parameter group of linear transformations, and consider the trajectory $\varphi: \mathbf{R} \to \mathbf{R}^n$ of a point $x_0 \in \mathbf{R}^n$.

Problem 1. Prove that $\varphi(t)$ is a solution of the equation

$$\dot{\mathbf{x}} = A\mathbf{x} \tag{1}$$

satisfying the initial condition $\varphi(0) = \mathbf{x}$, where $A: \mathbf{R}^n \to \mathbf{R}^n$ is the linear operator ($\equiv$ an $\mathbf{R}$-endomorphism) defined by the formula

$$A\mathbf{x} = \left.\frac{d}{dt}\right|_{t=0} g^t\mathbf{x} \quad \forall \mathbf{x} \in \mathbf{R}^n.$$

Hint. See Sec. 3.3.

Equation (1) is said to be *linear*. Thus, to describe all one-parameter groups of linear transformations, we need only investigate the solutions of the linear equation (1).

We will see later that the correspondence between one-parameter groups $\{g^t\}$ of linear transformations and linear equations of the type (1) is one-to-one. Thus every operator $A \colon \mathbf{R}^n \to \mathbf{R}^n$ specifies a one-parameter group $\{g^t\}$.

Example 1. Let $n = 1$, and let A be multiplication by the number k. Then g^t is an e^{kt}-fold expansion.

Problem 2. Find the velocity field of points of a rigid body rotating with angular velocity ω about an axis going through the origin.

13.3. Linear equations. Let $A \colon \mathbf{R}^n \to \mathbf{R}^n$ be a linear operator in the n-dimensional real space $\mathbf{R}^n$.

Definition. By a *linear equation* is meant an equation with phase space $\mathbf{R}^n$ determined by a velocity field $\mathbf{v}(\mathbf{x}) = A\mathbf{x}$:

$$\dot{\mathbf{x}} = A\mathbf{x}. \tag{1}$$

The full description of equation (1) is "a system of n homogeneous linear differential equations of the first order with constant real coefficients."

Let $x_i, i = 1, \ldots, n$ be a fixed system of (linear) coordinates in $\mathbf{R}^n$. Then equation (1) can be written as a system of n equations

$$\dot{x}_i = \sum_{j=1}^{n} a_{ij} x_j, \qquad i = 1, \ldots, n, \tag{1'}$$

where (a_{ij}) is the matrix of the operator A in the given coordinate system. This matrix is called the *matrix of the system* (1').

For $n = 1$ the solution of equation (1) satisfying the initial condition $\varphi(0) = \mathbf{x}_0$ is given by the exponential

$$\varphi(t) = e^{tA}\mathbf{x}_0.$$

It turns out that the solution is still given by the same formula in the general case, provided we explain what is meant by the exponential of a linear operator. We now turn our attention to this problem.

14. The Exponential of an Operator

The function e^A, $A \in \mathbf{R}$ can be defined in either of two equivalent ways:

$$e^A = E + A + \frac{A^2}{2!} + \frac{A^3}{3!} + \cdots, \tag{1}$$

$$e^A = \lim_{n \to \infty} \left(E + \frac{A}{n} \right)^n \tag{2}$$

(where E denotes unity).

Now let $A \colon \mathbf{R}^n \to \mathbf{R}^n$ be a linear operator. To define e^A, we must first define the concept of the limit of a sequence of linear operators.

14.1. The norm of an operator. Let $(\cdot, \cdot)$ be a scalar product in $\mathbf{R}^n$, and let $|\mathbf{x}| = \sqrt{(\mathbf{x}, \mathbf{x})}$ be the norm of the vector $\mathbf{x} \in \mathbf{R}^n$, i.e., the square root of the scalar product of $\mathbf{x}$ with itself.

Definition. By the *norm* of a linear operator $A: \mathbf{R}^n \to \mathbf{R}^n$ is meant the number

$$|A| = \sup_{\mathbf{x} \neq 0} \frac{|A\mathbf{x}|}{|\mathbf{x}|}.$$

Geometrically $|A|$ is just the largest "expansion coefficient" of the transformation A.

Problem 1. Prove that $0 \leqslant |A| < \infty$.

Hint. $|A| = \sup_{|\mathbf{x}| = 1} |A\mathbf{x}|$, the sphere is compact, and the function $|A\mathbf{x}|$ is continuous.

Problem 2. Prove that

$$|\lambda A| = |\lambda||A|, \quad |A + B| \leqslant |A| + |B|, \quad |AB| \leqslant |A||B|,$$

where $A, B: \mathbf{R}^n \to \mathbf{R}^n$ are linear operators and $\lambda \in \mathbf{R}$ is a number.

Problem 3. Let (a_{ij}) be the matrix of the operator A in an orthonormal basis. Prove that $\max_j \sum_i a_{ij}^2 \leqslant |A|^2 \leqslant \sum_{i,j} |a_{ij}|^2$.

Hint. See G. E. Shilov, *An Introduction to the Theory of Linear Spaces* (translated by R. A. Silverman), Dover, New York (1974), Sec. 53.

14.2. The metric space of operators. The set L of all linear operators $A: \mathbf{R}^n \to \mathbf{R}^n$ is itself a linear space over the field of real numbers (by definition, $(A + \lambda B)x = Ax + \lambda Bx$).

Problem 1. What is the dimension of the linear space L?

Ans. n^2.

Hint. An operator is specified by its matrix.

We now define the distance between two operators as the norm of the difference $A - B$:

$$\rho(A, B) = |A - B|. \tag{3}$$

THEOREM. *The space of linear operators with the metric ρ is a complete metric space.*†

† By a *metric space* is meant a pair consisting of a set M and a function $\rho: M \times M \to \mathbf{R}$, called the *metric*, such that
1) $\rho(x, y) \geqslant 0 \;\forall\; x, y \in M, \rho(x, y) = 0$ if and only if $x = y$;
2) $\rho(x, y) = \rho(y, x) \;\forall\; x, y \in M$;
3) $\rho(x, y) \leqslant \rho(x, z) + \rho(z, y) \;\forall\; x, y, z \in M$.
A sequence x_i of points of a metric space M is called a *Cauchy sequence* if $\forall\; \varepsilon > 0 \;\exists\; N: \rho(x_i, x_j) < \varepsilon \;\forall\; i, j > N$. A sequence x_i is said to *converge* to a point x if $\forall\; \varepsilon > 0 \;\exists\; N: \rho(x, x_i) < \varepsilon \;\forall\; i > N$. The space M is said to be *complete* if every Cauchy sequence is convergent.

Proof. It follows from the definition (3) that $\rho > 0$ if $A \neq B$, $\rho(A, A) = 0$, $\rho(B, A) = \rho(A, B)$. The triangle inequality

$$\rho(A, C) \leqslant \rho(A, B) + \rho(B, C)$$

is an immediate consequence of the inequality $|X + Y| \leqslant |X| + |Y|$ (Sec. 14.1, Problem 2) if we set $X = A - B$, $Y = B - C$. Thus ρ is a metric, and the space L equipped with ρ is a metric space. The completeness of L is easily proved (see below). ∎

14.3. Proof of completeness. Let A_i be a Cauchy sequence, i.e., suppose that for every $\varepsilon > 0$, there is an $N(\varepsilon) > 0$ such that $\rho(A_m, A_k) < \varepsilon$ if $m, k > N$. Given any $\mathbf{x} \in \mathbf{R}^n$, form the sequence of points $\mathbf{x}_i \in \mathbf{R}^n$, $\mathbf{x}_i = A_i \mathbf{x}$. Then $\mathbf{x}_i$ is a Cauchy sequence in the space $\mathbf{R}^n$ equipped with the Euclidean metric $\rho(\mathbf{x}, \mathbf{y}) = |\mathbf{x} - \mathbf{y}|$. In fact, by the definition of the norm of an operator,

$$|\mathbf{x}_m - \mathbf{x}_k| \leqslant \rho(A_m, A_k)|\mathbf{x}| \leqslant \varepsilon|\mathbf{x}|$$

for $m, k > N$. Since $|\mathbf{x}|$ is a fixed number (independent of m and k), it follows that $\mathbf{x}_i$ is a Cauchy sequence. The space $\mathbf{R}^n$ is complete, and hence the limit

$$\mathbf{y} = \lim_{i \to \infty} \mathbf{x}_i \in \mathbf{R}^n$$

exists. Note that $|\mathbf{x}_k - \mathbf{y}| \leqslant \varepsilon|\mathbf{x}|$ for $k > N(\varepsilon)$, where $N(\varepsilon)$ is the same number independent of $\mathbf{x}$ as above. The point $\mathbf{y}$ depends linearly on the point $\mathbf{x}$ (the limit of a sum equals the sum of the limits). This gives a linear operator $A: \mathbf{R}^n \to \mathbf{R}^n$, $A\mathbf{x} = \mathbf{y}$, $A \in L$. But

$$\rho(A_k, A) = |A_k - A| = \sup_{\mathbf{x} \neq 0} \frac{|\mathbf{x}_k - \mathbf{y}|}{|\mathbf{x}|} \leqslant \varepsilon$$

for $k > N(\varepsilon)$. Therefore

$$A = \lim_{k \to \infty} A_k,$$

and the space L is complete. ∎

Problem 1. Prove that a sequence of operators A_i converges if and only if the sequence of their matrices in a fixed basis converges. Use this to give another proof of completeness.

14.4. Series. Let M be a real linear space, provided with a metric ρ such that the distance between two points of M depends only on the difference between the points and

$$\rho(\lambda x, 0) = |\lambda|\rho(x, 0), \qquad x \in M, \qquad \lambda \in \mathbf{R}.$$

Suppose also that M, taken with this metric, is a complete metric space. Then M is said to be a *normed* linear space, and the function $\rho(x, 0)$ is called the *norm* of x and is denoted by $|x|$.

Example 1. Euclidean space $M = \mathbf{R}^n$ with the metric

$$\rho(x, y) = |\mathbf{x} - \mathbf{y}| = \sqrt{(\mathbf{x} - \mathbf{y}, \mathbf{x} - \mathbf{y})}.$$

Example 2. The space L of linear operators $A, B: \mathbf{R}^n \to \mathbf{R}^n$ with the metric

$$\rho(A, B) = |A - B|.$$

The distance between elements A, $B \in M$ will be denoted by $|A - B|$. Since the elements of M can be added and multiplied by numbers and since Cauchy sequences in M have limits, the theory of series of the form

$$A_1 + A_2 + \cdots, \qquad A_i \in M$$

is literally the same as the theory of numerical series. The theory of series of functions can also be carried over at once to the case of functions with values in M.

Problem 1. Prove the following two theorems:

WEIERSTRASS' TEST. *If the series*

$$\sum_{i=1}^{\infty} f_i \tag{4}$$

of functions $f_i\colon X \to M$ is majorized by a convergent numerical series, i.e., if

$$|f_i| < a_i, \qquad \sum_{i=1}^{\infty} a_i < \infty, \qquad a_i \in \mathbf{R},$$

then the series (4) *is absolutely and uniformly convergent on X.*

DIFFERENTIATION OF SERIES. *If the series* (4) *of functions $f_i\colon \mathbf{R} \to M$ is convergent and if the series of derivatives*

$$\sum_{i=1}^{\infty} \frac{df_i}{dt} \tag{5}$$

is uniformly convergent, then the series (4) *can be differentiated term by term (t is the coordinate on the line $\mathbf{R}$):*

$$\frac{d}{dt} \sum_{i=1}^{\infty} f_i = \sum_{i=1}^{\infty} \frac{df_i}{dt}.$$

Hint. The proof for the case $M = \mathbf{R}$ is given in advanced calculus and can be carried over word for word to the general case.

14.5. Definition of the exponential e^A. Let $A\colon \mathbf{R}^n \to \mathbf{R}^n$ be a linear operator.

Definition. By the *exponential e^A of the operator A* is meant the linear operator

$$e^A = E + A + \frac{A^2}{2!} + \cdots = \sum_{k=0}^{\infty} \frac{A^k}{k!},$$

where E is the identity operator ($Ex = x$).

THEOREM. *Given any A, the series e^A is uniformly convergent on every set $X = \{A\colon |A| \leqslant a\}$, $a \in \mathbf{R}$.*

Proof. If $|A| \leqslant a$, the series e^A is majorized by the numerical series

$$1 + a + \frac{a^2}{2!} + \cdots,$$

which converges to e^a. It follows from Weierstrass' test that the series e^A is uniformly convergent for $|A| \leqslant a$. ∎

Problem 1. Calculate the matrix e^{tA} if the matrix A is of the form

a) $\begin{pmatrix} 1 & 0 \\ 0 & 2 \end{pmatrix}$;   b) $\begin{pmatrix} 0 & 1 \\ 0 & 0 \end{pmatrix}$;   c) $\begin{pmatrix} 0 & 1 \\ -1 & 0 \end{pmatrix}$;   d) $\begin{pmatrix} 0 & 1 & 0 \\ 0 & 0 & 1 \\ 0 & 0 & 0 \end{pmatrix}$.

14.6. Example. Consider the set of all polynomials of degree less than n in a variable x with real coefficients. This set has the natural structure of a real linear space, since polynomials can be added and multiplied by numbers.

Problem 1. Find the dimension of the space of all polynomials of degree less than n.

Ans. n; for example, $1, x, x^2, \ldots, x^{n-1}$ is a basis.

We will denote the space of all polynomials of degree less than n by $\mathbf{R}^n$.† The derivative of a polynomial p of degree less than n is itself a polynomial of degree less than n. This gives rise to the mapping

$$A : \mathbf{R}^n \to \mathbf{R}^n, \qquad Ap = \frac{dp}{dx}. \tag{6}$$

Problem 2. Prove that A is a linear operator, and find its kernel and image.

Ans. Ker $A = \mathbf{R}^1$, Im $A = \mathbf{R}^{n-1}$.

On the other hand, let H^t $(t \in \mathbf{R})$ denote the operator of shift by t, carrying the polynomial $p(x)$ into $p(x + t)$.

Problem 3. Prove that $H^t : \mathbf{R}^n \to \mathbf{R}^n$ is a linear operator, and find its kernel and image.

Ans. Ker $H^t = 0$, Im $H^t = \mathbf{R}^n$.

Finally we form the operator e^{tA}.

THEOREM. *If A is the operator* (6), *then*

$$e^{tA} = H^t.$$

Proof. This is just Taylor's formula for polynomials

$$p(x + t) = p(x) + \frac{t}{1!}\frac{dp}{dx} + \frac{t^2}{2!}\frac{d^2p}{dx^2} + \cdots$$

(familiar from calculus). ∎

† Thus we identify the space of polynomials, equipped with the basis indicated in Problem 1, with the isomorphic coordinate space $\mathbf{R}^n$.

14.7. The exponential of a diagonal operator. Suppose the matrix of the operator A is diagonal, with diagonal elements $\lambda_1, \ldots, \lambda_n$. Then it is easy to see that the matrix of the operator e^A is also diagonal, with diagonal elements $e^{\lambda_1}, \ldots, e^{\lambda_n}$.

Definition. An operator $A: \mathbf{R}^n \to \mathbf{R}^n$ is said to be *diagonal* if its matrix is diagonal in some basis. Such a basis is called an *eigenbasis*.

Problem 1. Give an example of a nondiagonal operator.

Problem 2. Prove that the eigenvalues of a diagonal operator A are real.

Problem 3. Prove that if all n eigenvalues of an operator $A: \mathbf{R}^n \to \mathbf{R}^n$ are real and distinct, then A is diagonal.

   Let A be a diagonal operator. Then e^A is most easily calculated in an eigenbasis.

Example 1. Suppose the operator A has a matrix of the form

$$\begin{pmatrix} 1 & 1 \\ 1 & 1 \end{pmatrix}$$

in a basis $\mathbf{e}_1, \mathbf{e}_2$. Since the eigenvalues $\lambda_1 = 2, \lambda_2 = 0$ are real and distinct, the operator A is diagonal with eigenbasis $\mathbf{f}_1 = \mathbf{e}_1 + \mathbf{e}_2, \mathbf{f}_2 = \mathbf{e}_1 - \mathbf{e}_2$. The matrix of A in this basis is just

$$\begin{pmatrix} 2 & 0 \\ 0 & 0 \end{pmatrix}.$$

Hence the matrix of the operator e^A in the eigenbasis is

$$\begin{pmatrix} e^2 & 0 \\ 0 & 1 \end{pmatrix}.$$

Thus the matrix of the operator e^A is

$$\frac{1}{2}\begin{pmatrix} e^2 + 1 & e^2 - 1 \\ e^2 - 1 & e^2 + 1 \end{pmatrix}$$

in the original basis.

14.8. The exponential of a nilpotent operator.

Definition. An operator $A: \mathbf{R}^n \to \mathbf{R}^n$ is said to be *nilpotent* if some power of A equals 0.

Problem 1. Prove that the operator with matrix

$$\begin{pmatrix} 0 & 1 \\ 0 & 0 \end{pmatrix}$$

is nilpotent. More generally, prove that if all the elements of the matrix of an operator on and below the main diagonal are zero, then the operator is nilpotent.

Problem 2. Prove that the differentiation operator d/dx in the space of all polynomials of degree less than n is nilpotent.

If the operator A is nilpotent, then the series e^A terminates, i.e., reduces to a finite sum.

Problem 3. Calculate e^{tA} ($t \in \mathbf{R}$) where $A : \mathbf{R}^n \to \mathbf{R}^n$ is the operator with matrix

$$\begin{pmatrix} 0 & 1 & & & 0 \\ & 0 & \cdot & & \\ & & \cdot & \cdot & \\ & & & \cdot & 1 \\ 0 & & & \cdot & 0 \end{pmatrix}$$

(1 over the main diagonal and 0 elsewhere).

Hint. One way of solving this problem is to use Taylor's formula for polynomials. The differential operator d/dx has a matrix of the indicated type in some basis (which one?). For further details, see Sec. 25.

14.9. Quasi-polynomials. Let λ be a fixed real number. Then by a *quasi-polynomial with exponent λ* is meant a product of the form $e^{\lambda x}p(x)$ where p is a polynomial. The degree of p is called the *degree* of the quasi-polynomial.

Problem 1. Prove that the set of all quasi-polynomials with exponent λ of degree less than n is a linear space. What is the dimension of this space?

Ans. n; for example, $e^{\lambda x}, xe^{\lambda x}, \ldots, x^{n-1}e^{\lambda x}$ is a basis.

Remark. There is a certain ambiguity implicit in the concept of a quasi-polynomial, just as in the case of a polynomial. A (quasi-) polynomial can be regarded as an *expression* made up of signs and letters, in which case the solution of the preceding problem is obvious. On the other hand, we can regard a (quasi-) polynomial as a *function*, i.e., as a mapping $f : \mathbf{R} \to \mathbf{R}$. Actually both concepts are equivalent (when the coefficients of the polynomials are real or complex numbers†).

Problem 2. Prove that every function $f : \mathbf{R} \to \mathbf{R}$ which can be written as a quasi-polynomial has a unique representation as a quasi-polynomial.

Hint. We need only note that if $e^{\lambda x}p(x) \equiv 0$, then the coefficients of the polynomial $p(x)$ all vanish.

The n-dimensional linear space of quasi-polynomials of degree less than n with exponent λ will be denoted by $\mathbf{R}^n$.

THEOREM. *The differential operator d/dx is a linear operator from $\mathbf{R}^n$ to $\mathbf{R}^n$ such that*

$$e^{td/dx} = H^t \tag{7}$$

for every $t \in R$, where $H^t : \mathbf{R}^n \to \mathbf{R}^n$ is the operator of shift by t, i.e., $(H^t f)(x) = f(x + t)$.

Proof. Proving first that the derivative and shift of a quasi-polynomial of

---

† We will soon consider (quasi-) polynomials with real coefficients.

degree less than n is itself a quasi-polynomial of degree less than n, we note that

$$\frac{d}{dx}(e^{\lambda x}p(x)) = \lambda e^{\lambda x}p(x) + e^{\lambda x}p'(x),$$

$$e^{\lambda(x+t)}p(x + t) = e^{\lambda x}e^{\lambda t}p(x + t).$$

Moreover, the linearity of both the derivative and the shift is apparent. Note also that the Taylor series of a quasi-polynomial is absolutely convergent on the whole real line (since the Taylor series of $e^{\lambda x}$ and $p(x)$ are absolutely convergent). Comparing the Taylor series

$$f(x + t) = \sum_{n=0}^{\infty} \frac{f^{(n)}(x)}{n!}t^n$$

and the expansion

$$e^{tA} = \sum_{n=0}^{\infty} \frac{A^n}{n!}t^n,$$

we get (7). ∎

Problem 3. Calculate the matrix of the operator e^{tA} if the matrix of A is of the form

$$\begin{pmatrix} \lambda & 1 & & 0 \\ & \lambda & \ddots & \\ & & \ddots & 1 \\ 0 & & & \lambda \end{pmatrix}$$

(λ on the main diagonal, 1 over the main diagonal, 0 elsewhere). For example, calculate

$$\exp\begin{pmatrix} 1 & 1 \\ 0 & 1 \end{pmatrix}.$$

Hint. This is precisely the form of the matrix of the differentiation operator in the space of quasi-polynomials (in which basis?). For further details, see Sec. 25.

15. Properties of the Exponential

We now establish a number of properties of the operator $e^A : \mathbf{R}^n \to \mathbf{R}^n$. These properties allow us to use e^A to solve linear differential equations.

15.1. The group property. Let $A : \mathbf{R}^n \to \mathbf{R}^n$ be a linear operator.

THEOREM. *The family of linear operators $e^{tA} : \mathbf{R}^n \to \mathbf{R}^n$, $t \in \mathbf{R}$ is a one-parameter group of linear transformations of $\mathbf{R}^n$.*

Proof. Since it is already known that e^{tA} is a linear operator, we need only verify that

$$e^{(t+s)A} = e^{tA}e^{sA} \tag{1}$$

and that e^{tA} depends differentiably on t. In fact, we will show that

$$\frac{d}{dt}e^{tA} = Ae^{tA}, \tag{2}$$

as might be expected of an exponential. To prove the group property (1), we first multiply the series in powers of A formally, obtaining

$$\left(E + tA + \frac{t^2}{2}A^2 + \cdots\right)\left(E + sA + \frac{s^2}{2}A^2 + \cdots\right)$$
$$= E + (t + s)A + \left(\frac{t^2}{2} + ts + \frac{s^2}{2}\right)A^2 + \cdots.$$

The coefficient of A^k in the product equals $(t + s)^k/k!$, since formula (1) holds in the case of numerical series ($A \in \mathbf{R}$). The legitimacy of the term-by-term multiplication is proved in the same way as the legitimacy of the term-by-term multiplication of absolutely convergent numerical series (the series for e^{tA} and e^{sA} are absolutely convergent, since the series for $e^{|t|a}$ and $e^{|s|a}$ where $a = |A|$ are convergent).

To prove (2), we differentiate the series for e^{at} with respect to t formally, obtaining a series of derivatives:

$$\sum_{k=0}^{\infty}\frac{d}{dt}\frac{t^k}{k!}A^k = A\sum_{k=0}^{\infty}\frac{t^k}{k!}A^k.$$

This series converges absolutely and uniformly in any domain $|A| \leqslant a$, $|t| \leqslant T$, just like the original series. Hence the derivative of the sum of the series exists and equals the sum of the series of derivatives. ∎

We can also prove (1) by reducing the proof directly to the numerical case, after first proving the following

LEMMA. *Let $p \in \mathbf{R}[z_1, \ldots, z_N]$ be a polynomial in the variables $z_1, \ldots, z_N$ with nonnegative coefficients, and let $A_1, \ldots, A_N: \mathbf{R}^n \to \mathbf{R}^n$ be linear operators. Then*

$$|p(A_1, \ldots, A_N)| \leqslant p(|A_1|, \ldots, |A_N|).$$

Proof. An immediate consequence of Sec. 14.1, Problem 2. ∎

Proof of formula (1). Let $S_m(A)$ denote the partial sum of the series for e^A:

$$S_m(A) = \sum_{k=0}^{m}\frac{A^k}{k!}.$$

Then S_m is a polynomial in A with nonnegative coefficients. We must show that the difference

$$\Delta_m = S_m(tA)S_m(sA) - S_m((t + s)A)$$

converges to 0 as $m \to \infty$. Note that Δ_m is a polynomial in sA and tA with *nonnegative coefficients*. In fact, the terms in the product series of degree no higher than m in A are all obtained by multiplying the terms in the factor series of degree no higher than m in A.

Moreover, $S_m((s + t)A)$ is a partial sum of the product series, and hence Δ_m is the sum of all terms in the product $S_m(tA)S_m(sA)$ of degree higher than m in A. But all the coefficients of a product of polynomials with nonnegative coefficients are nonnegative.

It follows from the lemma that

$$|\Delta_m(tA, sA)| \leqslant \Delta_m(|tA|, |sA|).$$

Let τ and σ denote the nonnegative numbers $|tA|$ and $|sA|$, so that

$$\Delta_m(\tau, \sigma) = S_m(\tau)S_m(\sigma) - S_m(\tau + \sigma).$$

Since $e^\tau e^\sigma = e^{\tau + \sigma}$, the right-hand side approaches 0 as $m \to \infty$. Thus

$$\lim_{m \to \infty} \Delta_m(tA, sA) = 0,$$

and formula (1) is proved. ∎

Problem 1. Is it true that $e^{A+B} = e^A e^B$.

Ans. No.

Problem 2. Prove that $\det e^A \neq 0$.

Hint. $e^{-A} = (e^A)^{-1}$.

Problem 3. Prove that if A is an antisymmetric operator in Euclidean space, then the operator e^A is orthogonal.

15.2. The basic theorem of the theory of linear equations with constant coefficients.

Theorem 15.1 immediately implies a formula for the solution of the differential equation

$$\dot{\mathbf{x}} = A\mathbf{x}, \qquad \mathbf{x} \in \mathbf{R}^n. \tag{3}$$

THEOREM. *The solution of equation* (3) *satisfying the initial condition* $\boldsymbol{\varphi}(0) = \mathbf{x}_0$ *is*

$$\boldsymbol{\varphi}(t) = e^{tA}\mathbf{x}_0, \qquad t \in \mathbf{R}. \tag{4}$$

Proof. According to the differentiation formula (2),

$$\frac{d\boldsymbol{\varphi}}{dt} = Ae^{tA}\mathbf{x}_0 = A\boldsymbol{\varphi}(t),$$

so that $\boldsymbol{\varphi}$ is a solution. Moreover $e^0 = E$, $\boldsymbol{\varphi}(0) = \mathbf{x}_0$. This proves the theorem, since by the uniqueness theorem every solution coincides with (4) in its domain of definition. ∎

15.3. The general form of one-parameter groups of linear transformations of the space $\mathbf{R}^n$.

THEOREM. *Let* $g^t \colon \mathbf{R}^n \to \mathbf{R}^n$ *be a one-parameter group of linear transformations. Then there exists a linear operator* $A \colon \mathbf{R}^n \to \mathbf{R}^n$ *such that* $g^t = e^{tA}$.

Proof. Let

$$A = \frac{dg^t}{dt}\bigg|_{t=0} = \lim_{t \to 0} \frac{g^t - E}{t}.$$

We have already proved (see Sec. 3.2 and Problem 1, p. 96) that the trajectory $\varphi(t) = g^t\mathbf{x}_0$ is a solution of equation (3) satisfying the initial condition $\varphi(0) = \mathbf{x}_0$. But $g^t\mathbf{x}_0 = e^{tA}\mathbf{x}_0$ because of (4).   ∎

The operator A is called the *infinitesimal generator* of the group $\{g^t\}$.

Problem 1. Prove that the infinitesimal generator is uniquely determined by the group.

Remark. Thus there is a one-to-one correspondence between linear differential equations of the form (3) and their phase flows $\{g^t\}$, where each phase flow consists of linear diffeomorphisms.

15.4. Another definition of the exponential.

THEOREM. *If $A: \mathbf{R}^n \to \mathbf{R}^n$ is a linear operator, then*

$$e^A = \lim_{m \to \infty} \left(E + \frac{A}{m}\right)^m. \tag{5}$$

Proof. Consider the difference

$$e^A - \left(E + \frac{A}{m}\right)^m = \sum_{k=0}^{\infty} \left(\frac{1}{k!} - \frac{C_k^m}{m^k}\right)A^k$$

where the series converges since the series for e^A converges and

$$\left(E + \frac{A}{m}\right)^m$$

is a polynomial. The coefficients in the right-hand side are nonnegative since

$$\frac{1}{k!} \geqslant \frac{m(m-1)\cdots(m-k+1)}{m\cdot m\cdots m}\frac{1}{k!}.$$

Therefore, setting $|A| = a$, we get

$$\left|e^A - \left(E + \frac{A}{m}\right)^m\right| \leqslant \sum_{k=0}^{\infty}\left(\frac{1}{k!} - \frac{C_k^m}{m^k}\right)a^k = e^a - \left(1 + \frac{a}{m}\right)^m,$$

where the expression on the right approaches zero as $m \to \infty$.   ∎

15.5. Example: Euler's formula for e^z. Let $\mathbf{C}$ be the complex line. We
can regard $\mathbf{C}$ as the real plane $\mathbf{R}^2$ and multiplication by a complex number z as a linear operator $A: \mathbf{R}^2 \to \mathbf{R}^2$. The operator A is then a rotation through the angle arg z together with a $|z|$-fold expansion.

Problem 1. Find the matrix of multiplication by $z = u + iv$ in the basis $\mathbf{e}_1 = 1$, $\mathbf{e}_2 = i$.

Ans. $\begin{pmatrix} u & -v \\ v & u \end{pmatrix}$.

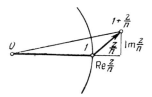

Fig. 103   The complex number $1 + (z/m)$.

We now find e^A. According to formula (5), we must first form the operator $E + (A/m)$ corresponding to multiplication by $1 + (z/m)$, i.e., rotation through the angle $\arg(1 + (z/m))$ together with expansion by a factor of $|1 + (z/m)|$ (Fig. 103).

Problem 2. Prove that

$$\arg\left(1 + \frac{z}{m}\right) = \operatorname{Im}\frac{z}{m} + o\left(\frac{1}{m}\right),$$

$$\left|1 + \frac{z}{m}\right| = 1 + \operatorname{Re}\frac{z}{m} + o\left(\frac{1}{m}\right) \tag{6}$$

as $m \to \infty$.

The operator $(E + (A/m))^m$ is a rotation through the angle $m \arg(1 + (z/m))$ together with an expansion by a factor of $|1 + (z/m)|^m$. Using (6), we find that the angle of rotation and the coefficient of expansion have the limiting values

$$\lim_{m \to \infty} m \arg\left(1 + \frac{z}{m}\right) = \operatorname{Im} z,$$

$$\lim_{m \to \infty} \left|1 + \frac{z}{m}\right|^m = e^{\operatorname{Re} z}. \tag{7}$$

THEOREM. *Let $z = u + iv$ be a complex number and $A: \mathbf{R}^2 \to \mathbf{R}^2$ the operator of multiplication by z. Then e^A is the operator of multiplication by the complex number $e^u(\cos v + i \sin v)$.*

Proof. An immediate consequence of (7). ∎

Definition. The complex number

$$e^u(\cos v + i \sin v) = \lim_{m \to \infty}\left(1 + \frac{z}{m}\right)^m$$

is called the *exponential* of the complex number $z = u + iv$ and is denoted by

$$e^z = e^u(\cos v + i \sin v). \tag{8}$$

Remark. If we identify the complex number z with the operation of multiplication by z, the definition reduces to a theorem, since the exponential of an operator has already been defined.

Problem 3. Find e^0, e^1, e^i, $e^{\pi i}$, $e^{2\pi i}$.

Problem 4. Prove that $e^{z_1 + z_2} = e^{z_1} e^{z_2}$ where z_1, $z_2 \in \mathbf{C}$.

Remark. Since the exponential is also defined by a series, we have

$$e^z = 1 + z + \frac{z^2}{2!} + \cdots, \qquad z \in \mathbf{C} \tag{9}$$

(the series is absolutely and uniformly convergent in every disk $|z| \leqslant a$).

Problem 5. Comparing this series with Euler's formula (8), deduce the Taylor series of $\sin v$ and $\cos v$.

Remark. Conversely, from a knowledge of the Taylor series of $\sin v$, $\cos v$, and e^u, we can prove formula (8), taking (9) as the definition of e^z.

15.6. Euler lines. Combining formulas (4) and (5), we get a method for approximate solution of the differential equation (3), known as the *method of Euler lines.*

Consider the differential equation with linear phase space $\mathbf{R}^n$ determined by a vector field $\mathbf{v}$. To find the solution $\boldsymbol{\varphi}$ of the equation $\dot{\mathbf{x}} = \mathbf{v}(\mathbf{x})$, $\mathbf{x} \in \mathbf{R}^n$ satisfying the initial condition $\mathbf{x}_0$, we proceed as follows (Fig. 104). The velocity at the point $\mathbf{x}_0$ is known and is just $\mathbf{v}(\mathbf{x}_0)$. Suppose we leave $\mathbf{x}_0$ and move with velocity $\mathbf{v}(\mathbf{x}_0)$ for a time interval $\Delta t = t/N$. Then we arrive at the point $\mathbf{x}_1 = \mathbf{x}_0 + \mathbf{v}(\mathbf{x}_0)\Delta t$. We then move with velocity $\mathbf{v}(\mathbf{x}_1)$ for another time interval Δt, and so on:

$$\mathbf{x}_{k+1} = \mathbf{x}_k + \mathbf{v}(\mathbf{x}_k)\Delta t, \qquad k = 0, 1, \ldots, N - 1.$$

The last point $\mathbf{x}_k$ will be denoted by $\mathbf{X}_N(t)$. Note that the graph representing the motion with piecewise-constant velocity is a polygonal curve (line) con-

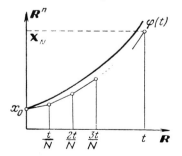

Fig. 104   An Euler line.

sisting of N segments in the extended phase space $\mathbf{R} \times \mathbf{R}^n$. This polygonal curve is known as an *Euler line*. It is natural to expect that as $N \to \infty$ the sequence of Euler lines will converge to an integral curve, so that for large N the last point $\mathbf{X}_N(t)$ will be close to the value of the solution $\boldsymbol{\varphi}$ at time t satisfying the initial condition $\boldsymbol{\varphi}(0) = \mathbf{x}_0$.

THEOREM. *For the linear equation* (3),

$$\lim_{N \to \infty} \mathbf{X}_N(t) = \boldsymbol{\varphi}(t). \tag{10}$$

Proof. It follows from the construction of the Euler line for $\mathbf{v}(\mathbf{x}) = A\mathbf{x}$ that

$$\mathbf{X}_N = \left(E + \frac{At}{N}\right)^N \mathbf{x}_0.$$

Therefore

$$\lim_{N \to \infty} \mathbf{X}_N(t) = e^{tA}\mathbf{x}_0,$$

by (5), which implies (10), by (4). ∎

Problem 1. Prove that not only does the end point of the Euler line approach $\boldsymbol{\varphi}(t)$, but also the whole sequence of piecewise-linear functions $\boldsymbol{\varphi}_n : I \to \mathbf{R}^n$, with the Euler lines as their graphs, converges uniformly to the solution $\boldsymbol{\varphi}$ on the interval $[0, t]$.

Remark. In the general case (where the vector field $\mathbf{v}$ depends on $\mathbf{x}$ *non-linearly*), the Euler line can also be written in the form

$$\mathbf{X}_N = \left(E + \frac{tA}{N}\right)^N \mathbf{x}_0,$$

where A is the nonlinear operator carrying the point $\mathbf{x}$ into the point $\mathbf{v}(\mathbf{x})$. We shall see later (Sec. 31.9) that even in this case the sequence of Euler lines converges to a solution, at least for sufficiently small $|t|$. Thus the expression (4), in which the exponential is defined by formula (5), gives the solution of all differential equations quite generally.†

The Eulerian theory of the exponential (which is essentially the same in all its variants), from the definition of the number e and the Euler and Taylor formulas for e^z up to formula (4) for the solution of linear equations and the method of Euler lines, has many other applications going beyond the scope of this course.

---

† In practice, the use of Euler lines is not a convenient way of solving differential equations approximately, since to obtain high accuracy we must choose a very small value of the "step" Δt. More often one uses various refinements of the Euler method, in which the integral curve is approximated not by a line segment, but rather by an arc of a parabola of some degree or other. The most frequently used methods are those of Adams, Störmer, and Runge, discussed in books on approximate computations.

16. The Determinant of the Exponential

Suppose the operator A is specified by its matrix. Then lengthy calculations may be required to find the matrix of the operator e^A. However, as we will soon see, the determinant of the matrix of e^A can be calculated very easily.

16.1. The determinant of an operator.

Definition. By the *determinant of a linear operator* $A: \mathbf{R}^n \to \mathbf{R}^n$, denoted by det A, is meant the determinant of the matrix of A in any basis $\mathbf{e}_1, \ldots, \mathbf{e}_n$.

The determinant of the matrix of the operator A does not depend on the basis. In fact, if (A) is the matrix of the operator A in the basis $\mathbf{e}_1, \ldots, \mathbf{e}_n$, then the matrix of A in another basis is of the form $(B)(A)(B^{-1})$. But clearly

$$\det (B)(A)(B^{-1}) = \det (A).$$

The determinant of a matrix is the oriented volume of the parallelepiped† whose *edges are given by the columns of the matrix.*

For example, for $n = 2$ (Fig. 105) the determinant

$$\begin{vmatrix} x_1 & x_2 \\ y_1 & y_2 \end{vmatrix}$$

is the area of the parallelogram spanned by the vectors $\boldsymbol{\xi}_1 = (x_1, y_1)$ and $\boldsymbol{\xi}_2 = (x_2, y_2)$, taken with the plus sign if the ordered pair of vectors $(\boldsymbol{\xi}_1, \boldsymbol{\xi}_2)$ specifies the same orientation of $\mathbf{R}^2$ as the pair of basis vectors $(\mathbf{e}_1, \mathbf{e}_2)$ and with the minus sign otherwise.

The ith column in the matrix of the operator A in the basis $\mathbf{e}_1, \ldots, \mathbf{e}_n$ is

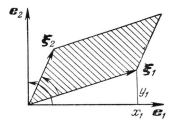

Fig. 105   The determinant of a matrix equals the oriented area of the parallelogram spanned by the columns of the matrix.

† The *parallelepiped* with edges $\boldsymbol{\xi}_1, \ldots, \boldsymbol{\xi}_n \in \mathbf{R}^n$ is the subset of $\mathbf{R}^n$ consisting of all points of the form $x_1\boldsymbol{\xi}_1 + \cdots + x_n\boldsymbol{\xi}_n$, $0 \leqslant x_i \leqslant 1$, $i = 1, 2, \ldots, n$. For $n = 2$ the parallelepiped is called a *parallelogram*. Starting from any definition of volume, we can easily prove the italicized assertion. Otherwise the assertion can be taken as the *definition* of the volume of a parallelepiped.

made up of the components of the image $A\mathbf{e}_i$ of the ith basis vector $\mathbf{e}_i$. Hence *the determinant of the operator A is the oriented volume of the image of the unit cube (the parallelepiped with edges $\mathbf{e}_1, \ldots, \mathbf{e}_n$) under the mapping A.*

Problem 1. Let Π be a parallelepiped with linearly independent edges. Prove that the ratio of the (oriented) volume of the image $A\Pi$ of the parallelepiped under the mapping A to the (oriented) volume of Π is independent of Π and equals det A.

Remark. The reader familiar with the theory of measurement of volumes in $\mathbf{R}^n$ will note that Π can be replaced by any other figure with volume.

Thus *the determinant of an operator A is the coefficient of expansion of oriented volume in the sense that the oriented volume of any figure is expanded by a factor of* det A *under application of A.* Geometrically, it is far from obvious that the volume expansion is the same for all figures (even in the planar case), since a linear transformation can drastically change the shape of a figure.

16.2. The trace of an operator. By the *trace* of a matrix $A = (a_{ij})$, denoted by Tr A,† is meant the sum of its diagonal elements

$$\mathrm{Tr}\, A = \sum_{i=1}^{n} a_{ii}.$$

The trace of the matrix of an operator $A\colon \mathbf{R}^n \to \mathbf{R}^n$ does not depend on the basis, but only on the operator itself.

Problem 1. Prove that the trace of a matrix equals the sum of all n of its eigenvalues, while the determinant equals the product of the eigenvalues.

Hint. Apply the formula

$$(\lambda - x_1) \cdots (\lambda - x_n) = \lambda^n - (x_1 + \cdots + x_n)\lambda^{n-1} + \cdots + (-1)^n x_1 \cdots x_n$$

to the polynomial

$$\det (A - \lambda E) = (-\lambda)^n + (-\lambda)^{n-1} \sum_{i=1}^{n} a_{ii} + \cdots.$$

Since the eigenvalues are independent of the basis, we have the following

Definition. By the *trace of an operator A* is meant the trace of its matrix in any (and hence in every) basis.

16.3. Relation between the determinant and the trace.

THEOREM. *If $A\colon \mathbf{R}^n \to \mathbf{R}^n$ is a linear operator and ε a real number, then*

$$\det(E + \varepsilon A) = 1 + \varepsilon\, \mathrm{Tr}\, A + O(\varepsilon^2)$$

as $\varepsilon \to 0$.

† The trace of A is sometimes denoted by Sp A (from the German word "Spur").

First proof. The determinant of the operator $E + \varepsilon A$ equals the product of the eigenvalues of the operator. But the eigenvalues of $E + \varepsilon A$ (with due regard for multiplicity) equal $1 + \varepsilon \lambda_i$, where the λ_i are the eigenvalues of A. It follows that

$$\det (E + \varepsilon A) = \sum_{i=1}^{n} (1 + \varepsilon \lambda_i) = 1 + \varepsilon \sum_{i=1}^{n} \lambda_i + O(\varepsilon^2). \quad \blacksquare$$

Second proof. Clearly $\varphi(\varepsilon) = \det(E + \varepsilon A)$ is a polynomial in ε such that $\varphi(0) = 1$. We must show that $\varphi'(0) = \operatorname{Tr} A$. Denoting the elements of the matrix $E + \varepsilon A$ by x_{ij}, we have

$$\frac{d\varphi}{d\varepsilon}\bigg|_{\varepsilon=0} = \sum_{i,j=1}^{n} \frac{\partial \Delta}{\partial x_{ij}}\bigg|_E \frac{dx_{ij}}{d\varepsilon},$$

where Δ is the determinant of $E + \varepsilon A = (x_{ij})$. By definition, the partial derivative $\partial \Delta / \partial x_{ij}|_E$ equals

$$\frac{d}{dh}\bigg|_{h=0} \det (E + h e_{ij}),$$

where (e_{ij}) is the matrix whose only nonzero element is a 1 in the ith row and jth column. But

$$\det (E + h e_{ij}) = \begin{cases} 1 & \text{if } i \neq j, \\ 1 + h & \text{if } i = j, \end{cases}$$

and hence

$$\frac{\partial \Delta}{\partial x_{ij}}\bigg|_E = \begin{cases} 0 & \text{if } i \neq j, \\ 1 & \text{if } i = j. \end{cases}$$

It follows that

$$\frac{d\varphi}{d\varepsilon}\bigg|_{\varepsilon=0} = \sum_{i=1}^{n} \frac{dx_{ii}}{d\varepsilon} = \sum_{i=1}^{n} a_{ii} = \operatorname{Tr} A. \quad \blacksquare$$

Incidentally, we have again proved that the trace is independent of the basis.

COROLLARY. *Suppose small changes are made in the edges of a parallelepiped. Then the main contribution to the change in volume of the parallelepiped is due to the change of each edge in its own direction, changes in the direction of the other edges making only a second-order contribution to the change in volume.*

For example, the area of the parallelogram shown in Fig. 106, which is close to being a square, differs from the area of the shaded rectangle only by infinitesimals of the second order.

This corollary can also be deduced from elementary geometrical considerations, leading to a purely geometric proof of the above theorem.

16.4. The determinant of the operator e^A.

THEOREM. *For any linear operator $A \colon \mathbf{R}^n \to \mathbf{R}^n$,*

$$\det e^A = e^{\operatorname{Tr} A}.$$

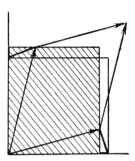

Fig. 106   Approximate determination of the area of a parallelogram which is close to being a square.

Proof. According to the second definition of the exponential,

$$\det e^A = \det \lim_{m \to \infty} \left(E + \frac{A}{m} \right)^m = \lim_{m \to \infty} \det \left(E + \frac{A}{m} \right)^m,$$

since the determinant of a matrix is a polynomial (and hence continuous) in its elements. Moreover, by Theorem 16.3,

$$\det \left(E + \frac{A}{m} \right)^m = \left[\det \left(E + \frac{A}{m} \right) \right]^m = \left[1 + \frac{1}{m} \operatorname{Tr} A + O \left(\frac{1}{m^2} \right) \right]^m, \quad m \to \infty.$$

It only remains to note that

$$\lim_{m \to \infty} \left[1 + \frac{a}{m} + O \left(\frac{1}{m^2} \right) \right]^m = e^a$$

for any $a \in \mathbf{R}$, in particular for $a = \operatorname{Tr} A$. ∎

COROLLARY 1. *The operator e^A is nonsingular.*

COROLLARY 2. *The operator e^A preserves the orientation of $\mathbf{R}^n$ (i.e., $\det e^A > 0$).*

COROLLARY 3 (**Liouville's formula**). *The t-advance mapping g^t of the linear equation*

$$\dot{\mathbf{x}} = A\mathbf{x}, \qquad \mathbf{x} \in \mathbf{R}^n \tag{1}$$

multiplies the volume of any figure by the factor e^{at}, where $a = \operatorname{Tr} A$.

Proof. Note that

$$\det g^t = \det e^{tA} = e^{\operatorname{Tr} tA} = e^{t \operatorname{Tr} A}. ∎$$

In particular, this implies

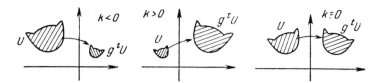

Fig. 107   Behavior of area under transformations of the phase flow of the pendulum equation with coefficient of friction $-k$.

COROLLARY 4. *If the trace of A equals 0, then the phase flow of equation (1) preserves volume, i.e., g^t carries every parallelepiped into a parallelepiped of equal volume.*

Proof. Merely note that $e^0 = 1$.   ∎

Example 1. Consider the equation

$$\ddot{x} = -x + k\dot{x}$$

of a pendulum with coefficient of friction $-k$, equivalent to the system

$$\begin{cases} \dot{x}_1 = x_2, \\ \dot{x}_2 = -x_1 + kx_2 \end{cases}$$

with matrix

$$\begin{pmatrix} 0 & 1 \\ -1 & k \end{pmatrix}$$

(Fig. 107). The trace of this matrix equals k. Let $\{g^t\}$ be the phase flow defined by the above system. Then if $k < 0$ the transformation g^t carries every domain of the phase plane into a domain of smaller area. On the other hand, in a system with negative friction $(k > 0)$, the area of the domain $g^t U, t > 0$ is larger than that of U. Finally, if there is no friction $(k = 0)$, the phase flow preserves area. This is hardly surprising, since in this last case, as we know from Sec. 6.6, g^t is a rotation through the angle t.

Problem 1. Suppose the real parts of all the eigenvalues of A are negative. Show that the transformations g^t of the phase flow of equation (1) then decrease volume $(t > 0)$.

Problem 2. Prove that the eigenvalues of the operator e^A equal e^{λ_i}, where the λ_i are the eigenvalues of the operator A. Use this to prove Theorem 16.4.

17. The Case of Distinct Real Eigenvalues

In practical problems involving differential equations, the matrix of the operator A is given in some basis and we must explicitly calculate the matrix of the operator e^A in the same basis. We begin by solving this problem in the particularly simple case where A has distinct real eigenvalues.

17.1. Diagonal operators. Consider the linear differential equation

$$\dot{\mathbf{x}} = A\mathbf{x}, \qquad \mathbf{x} \in \mathbf{R}^n, \tag{1}$$

where $A: \mathbf{R}^n \to \mathbf{R}^n$ is a diagonal operator. The matrix of the operator A is of the form

$$\begin{pmatrix} \lambda_1 & & 0 \\ & \ddots & \\ 0 & & \lambda_n \end{pmatrix}$$

in its eigenbasis,† where the λ_i are the eigenvalues of A. The matrix of the operator e^A has the form

$$\begin{pmatrix} e^{\lambda_1 t} & & 0 \\ & \ddots & \\ 0 & & e^{\lambda_n t} \end{pmatrix}$$

in the same basis. Thus the solution $\boldsymbol{\varphi}$ of equation (1) satisfying the initial condition $\boldsymbol{\varphi}(0) = (x_{10}, \ldots, x_{n0})$ has components

$$\varphi_k = e^{\lambda_k t} x_{k0}, \qquad k = 1, \ldots, n$$

in this basis.

If the n eigenvectors of the operator A are real and distinct, then A is diagonal ($\mathbf{R}^n$ decomposes into a direct sum of one-dimensional subspaces invariant under A). The procedure for solving (1) in this case goes as follows:

1) Form the *characteristic* (or *secular*) *equation*

$$\det (A - \lambda E) = 0;$$

2) Find the roots $\lambda_1, \ldots, \lambda_n$ of this equation (the λ_i are assumed to be real and distinct);

3) Find the eigenvectors $\boldsymbol{\xi}_1, \ldots, \boldsymbol{\xi}_n$ satisfying the linear equations

$$A\boldsymbol{\xi}_k = \lambda_k \boldsymbol{\xi}_k, \qquad \boldsymbol{\xi}_k \neq 0, \quad k = 1, \ldots, n;$$

4) Expand the initial condition with respect to the eigenvectors:

$$\mathbf{x}_0 = \sum_{k=1}^{n} C_k \boldsymbol{\xi}_k;$$

---

† We first go over to an eigenbasis if the matrix of the operator A is originally given in another basis.

5) Write the answer

$$\varphi(t) \; = \; \sum_{k=1}^{n} C_k e^{\lambda_k t} \xi_k.$$

In particular, we have the following

COROLLARY. *Let A be a diagonal operator. Then the elements of the matrix e^{tA} ($t \in \mathbf{R}$) in any basis are linear combinations of the exponentials $e^{\lambda_k t}$, where the λ_k are the eigenvalues of the matrix A.*

17.2. Example. Consider the pendulum with friction

$$\begin{cases} \dot{x}_1 = x_2, \\ \dot{x}_2 = -x_1 - kx_2. \end{cases}$$

The matrix of the operator A is then

$$\begin{pmatrix} 0 & 1 \\ -1 & -k \end{pmatrix},$$

so that

$$\operatorname{Tr} A = -k, \qquad \det A = 1.$$

The corresponding characteristic equation

$$\lambda^2 + k\lambda + 1 = 0$$

has distinct real roots if its discriminant is positive, i.e., if $|k| > 2$. Thus the operator A is diagonal if the coefficient of friction k is sufficiently large (in absolute value).

Now suppose $k > 2$. Then both roots λ_1, λ_2 are negative, and the equation takes the form

$$\begin{cases} \dot{y}_1 = \lambda_1 y_1, & \lambda_1 < 0, \\ \dot{y}_2 = \lambda_2 y_2, & \lambda_2 < 0 \end{cases}$$

in the eigenbasis. Therefore, as in Sec. 4, we get the solution

$$y_1(t) = e^{\lambda_1 t} y_1(0),$$
$$y_2(t) = e^{\lambda_2 t} y_2(0),$$

and the phase curves have a node as in Fig. 108. As $t \to +\infty$ all the solutions approach 0, and almost all the integral curves become tangent to the y_1-axis if $|\lambda_2| > |\lambda_1|$ (y_2 then approaches 0 faster than y_1). The picture in the plane (x_1, x_2) is obtained from that in the plane (y_1, y_2) by making a linear transformation.

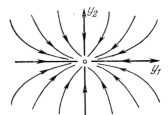

Fig. 108   Phase curves of the pendulum equation with strong friction in the eigenbasis.

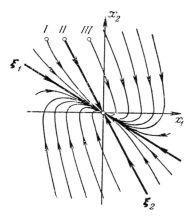

Fig. 109   Phase curves of the pendulum equation with strong friction in the usual basis.

For example, suppose $k = 10/3$, so that $\lambda_1 = -1/3$, $\lambda_2 = -3$. To find the eigenvector $\boldsymbol{\xi}_1$, we use the condition $x_1 = -3x_2$, obtaining $\boldsymbol{\xi}_1 = \mathbf{e}_2 - 3\mathbf{e}_1$. Similarly, we get $\boldsymbol{\xi}_2 = \mathbf{e}_1 - 3\mathbf{e}_2$. Since $|\lambda_1| < |\lambda_2|$, the phase curves have the form shown in Fig. 109. Studying Fig. 109, we come to the following remarkable conclusion: If the coefficient of friction k is sufficiently large $(k > 2)$, the pendulum does not execute damped oscillations, but rather goes directly into its equilibrium position; in fact, its velocity x_2 changes sign no more than once.

Problem 1. Find the motions of the pendulum corresponding to the phase curves I, II, and III in Fig. 109. Draw a typical graph of $x(t)$.

Problem 2. Investigate the motion of the inverted pendulum with friction:

$$\ddot{x} = x - k\dot{x}.$$

17.3. The discrete case. All that has been said about the exponential e^{tA} with a continuous argument t applies equally well to the exponential A^n with the discrete argument n. In particular, if A is a diagonal operator, A^n is most conveniently calculated by going over to a diagonal basis.

Problem 1. The Fibonacci sequence

$$0, 1, 1, 2, 3, 5, 8, 13, \ldots$$

is defined by the conditions

$$x_0 = 0, \qquad x_1 = 1, \qquad x_n = x_{n-1} + x_{n-2} \qquad \text{if} \qquad n \geqslant 2.$$

Find an explicit formula for x_n. Prove that x_n grows like a geometric progression, and find

$$\lim_{n \to \infty} \frac{\lim x_n}{n} = \alpha.$$

Hint. Note that the vector $\boldsymbol{\xi}_n = (x_n, x_{n-1})$ can be expressed linearly in terms of $\boldsymbol{\xi}_{n-1}$:

$$\boldsymbol{\xi}_1 = (1, 0), \qquad \boldsymbol{\xi}_n = A\boldsymbol{\xi}_{n-1}, \qquad A = \begin{pmatrix} 1 & 1 \\ 1 & 0 \end{pmatrix}.$$

Therefore x_n is the first component of the vector $A^{n-1}\boldsymbol{\xi}_1$.

Ans. $\alpha = \ln \dfrac{\sqrt{5} + 1}{2}$, $x_n = \dfrac{1}{\sqrt{5}}(\lambda_1^n - \lambda_2^n)$, where $\lambda_{1,2} = \dfrac{1 \pm \sqrt{5}}{2}$ are the eigenvalues of A.

Comment. The same argument reduces the study of any *recurrent sequence of order k*, defined by a relation

$$x_n = a_1 x_{n-1} + a_2 x_{n-2} + \cdots + a_k x_{n-k}, \qquad n \geqslant k,$$

together with the first k terms $x_0, x_1, \ldots, x_{k-1}$,[†] to the study of the exponential function A^n, where $A: \mathbf{R}^k \to \mathbf{R}^k$ is a linear operator. Therefore knowing how to calculate the matrix of an exponential enables us to calculate all recurrent sequences.

Returning to the general problem of calculating e^{tA}, we note that the roots of the characteristic equation $\det(A - \lambda E) = 0$ may be complex. To study this case, we first consider linear equations with a complex phase space $\mathbf{C}^n$.

18. Complexification and Decomplexification

Before studying complex differential equations, we introduce the concepts of complexification of a real space and decomplexification of a complex space.

18.1. Decomplexification. Let $\mathbf{C}^n$ denote an n-dimensional linear space over the field of complex numbers $\mathbf{C}$. Then by the *decomplexification of the space* $\mathbf{C}^n$ is meant the real linear space which coincides with $\mathbf{C}^n$ as a group and in which multiplication by real numbers is defined in the same way as

---

† The fact that the definition of a recurrent sequence of order k requires knowledge of the first k terms of the sequence is intimately connected with the fact that the phase space of a differential equation of order k is of dimension k. This connection becomes apparent if the differential equation is written as a limit of difference equations.

in $\mathbf{C}^n$, while multiplication by complex numbers is not defined at all. (In other words, to decomplexify $\mathbf{C}^n$ means to forget about the structure of the $\mathbf{C}$-module while preserving the structure of the $\mathbf{R}$-module.)

It is easy to see that the decomplexification of the space $\mathbf{C}^n$ is a $2n$-dimensional real linear space $\mathbf{R}^{2n}$. We will denote decomplexification by a superscript $\mathbf{R}$ on the left. Thus, for example, $^{\mathbf{R}}\mathbf{C} = \mathbf{R}^2$.

If $\mathbf{e}_1, \ldots, \mathbf{e}_n$ is a basis in $\mathbf{C}^n$, then $\mathbf{e}_1, \ldots, \mathbf{e}_n, i\mathbf{e}_1, \ldots, i\mathbf{e}_n$ is a basis in $^{\mathbf{R}}\mathbf{C}^n = \mathbf{R}^{2n}$.

Let $A: \mathbf{C}^m \to \mathbf{C}^n$ be a $\mathbf{C}$-linear operator. Then by the *decomplexification of the operator A* is meant the $\mathbf{R}$-linear operator $^{\mathbf{R}}A: {}^{\mathbf{R}}\mathbf{C}^m \to {}^{\mathbf{R}}\mathbf{C}^n$ which coincides with A pointwise.

Problem 1. Let $\mathbf{e}_1, \ldots, \mathbf{e}_m$ and $\mathbf{f}_1, \ldots, \mathbf{f}_n$ be bases in the spaces $\mathbf{C}^m$ and $\mathbf{C}^n$ respectively, and let (A) be the matrix of the operator A. Find the matrix of the decomplexified operator $^{\mathbf{R}}A$.

Ans. $\begin{pmatrix} \alpha & -\beta \\ \beta & \alpha \end{pmatrix}$, where $(A) = (\alpha) + i(\beta)$.

Problem 2. Prove that

$$^{\mathbf{R}}(A + B) = {}^{\mathbf{R}}A + {}^{\mathbf{R}}B, \qquad {}^{\mathbf{R}}(AB) = {}^{\mathbf{R}}A{}^{\mathbf{R}}B.$$

18.2. Complexification. Let $\mathbf{R}^n$ be an n-dimensional real linear space. Then by the *complexification of the space $\mathbf{R}^n$* is meant the n-dimensional complex linear space, denoted by $^{\mathbf{C}}\mathbf{R}^n$, which is constructed as follows. The points of $^{\mathbf{C}}\mathbf{R}^n$ are pairs (ξ, η) with $\xi \in \mathbf{R}^n$, $\eta \in \mathbf{R}^n$. Denoting such pairs (ξ, η) by $\xi + i\eta$, we define the operations of addition and multiplication by complex numbers in the usual way:

$$(\xi_1 + i\eta_1) + (\xi_2 + i\eta_2) = (\xi_1 + \xi_2) + i(\eta_1 + \eta_2),$$
$$(u + iv)(\xi + i\eta) = (u\xi - v\eta) + i(v\xi + u\eta).$$

It is easily verified that the resulting $\mathbf{C}$-module is an n-dimensional complex linear space $^{\mathbf{C}}\mathbf{R}^n = \mathbf{C}^n$. If $\mathbf{e}_1, \ldots, \mathbf{e}_n$ is a basis in $\mathbf{R}^n$, then the vectors $\mathbf{e}_1 + i\mathbf{0}, \ldots, \mathbf{e}_n + i\mathbf{0}$ form a $\mathbf{C}$-basis in $\mathbf{C}^n = {}^{\mathbf{C}}\mathbf{R}^n$. The vectors $\xi + i\mathbf{0}$ are denoted briefly by ξ.

Let $A: \mathbf{R}^m \to \mathbf{R}^n$ be an $\mathbf{R}$-linear operator. Then by the *complexification of the operator A* is meant the $\mathbf{C}$-linear operator $^{\mathbf{C}}A: {}^{\mathbf{C}}\mathbf{R}^m \to {}^{\mathbf{C}}\mathbf{R}^n$ defined by the formula

$$A(\xi + i\eta) = A\xi + iA\eta.$$

Problem 1. Let $\mathbf{e}_1, \ldots, \mathbf{e}_m$ and $\mathbf{f}_1, \ldots, \mathbf{f}_n$ be bases in the spaces $\mathbf{R}^m$ and $\mathbf{R}^n$ respectively, and let (A) be the matrix of the operator A. Find the matrix of the complexified operator $^{\mathbf{C}}A$.

Ans. $(^{\mathbf{C}}A) = (A)$.

Problem 2. Prove that

$$\mathbf{C}(A + B) = \mathbf{C}A + \mathbf{C}B, \qquad \mathbf{C}(AB) = \mathbf{C}A\mathbf{C}B.$$

Remark on terminology. The operations of complexification and decomplexification are defined both for spaces and for mappings. Algebraists call such operations *functors*.

18.3. The complex conjugate. Consider the $2n$-dimensional real linear space $\mathbf{R}^{2n} = {}^{\mathbf{R}\mathbf{C}}\mathbf{R}^n$, obtained from $\mathbf{R}^n$ by complexification followed by decomplexification. This space contains an n-dimensional subspace of vectors of the form $\xi + i\mathbf{0}$, $\xi \in \mathbf{R}^n$, called the *real plane* $\mathbf{R}^n \subset \mathbf{R}^{2n}$. The subspace of vectors of the form $\mathbf{0} + i\xi$, $\xi \in \mathbf{R}^n$ is called the *imaginary plane* $i\mathbf{R}^n \subset \mathbf{R}^{2n}$. The whole space $\mathbf{R}^{2n}$ is the direct sum of these two n-dimensional subspaces.

The operator iE of multiplication by i in $\mathbf{C}^n = {}^{\mathbf{C}}\mathbf{R}^n$ is transformed after decomplexification into an $\mathbf{R}$-linear operator ${}^{\mathbf{R}}(iE) = I: \mathbf{R}^{2n} \to \mathbf{R}^{2n}$ (Fig. 110). This operator I maps the real plane isomorphically into the imaginary plane and vice versa. The square of the operator I equals $-E$.

Problem 1. Let $\mathbf{e}_1, \ldots, \mathbf{e}_n$ be a basis in $\mathbf{R}^n$ and $\mathbf{e}_1, \ldots, \mathbf{e}_n, i\mathbf{e}_1, \ldots, i\mathbf{e}_n$ a basis in $\mathbf{R}^{2n} = {}^{\mathbf{R}\mathbf{C}}\mathbf{R}^n$. Find the matrix of the operator I in this basis.

Ans. $(I) = \begin{pmatrix} 0 & -E \\ E & 0 \end{pmatrix}.$

Let $\sigma: \mathbf{R}^{2n} \to \mathbf{R}^{2n}$ (Fig. 111) denote the operator of taking the complex conjugate, so that $\sigma(\xi + i\eta) = \xi - i\eta$. The action of σ is often denoted by an overbar. The operator σ coincides with E in the real plane and with $-E$ in the imaginary plane. Note that σ is *involutory*: $\sigma^2 = E$.

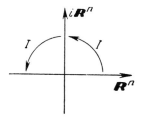

Fig. 110 The operator of multiplication by i.

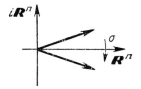

Fig. 111 The complex conjugate.

Let $A\colon {}^{\mathbf{C}}\mathbf{R}^m \to {}^{\mathbf{C}}\mathbf{R}^n$ be a **C**-linear operator. By the *complex conjugate $\bar{A}$ of the operator A* is meant the operator $\bar{A}\colon {}^{\mathbf{C}}\mathbf{R}^m \to {}^{\mathbf{C}}\mathbf{R}^n$ defined by the formula

$$\overline{A\mathbf{z}} = \bar{A}\,\bar{\mathbf{z}} \quad \forall\, \mathbf{z} \in {}^{\mathbf{C}}\mathbf{R}^m.$$

Problem 2. Prove that $\bar{A}$ is a **C**-linear operator.

Problem 3. Prove that the matrix of the operator $\bar{A}$ *in a real basis* is the complex conjugate of the matrix of A in the same basis.

Problem 4. Prove that

$$\overline{A + B} = \bar{A} + \bar{B}, \qquad \overline{AB} = \bar{A}\,\bar{B}, \qquad \overline{\lambda A} = \bar{\lambda}\,\bar{A}.$$

Problem 5. Prove that a complex linear operator $A\colon {}^{\mathbf{C}}\mathbf{R}^m \to {}^{\mathbf{C}}\mathbf{R}^n$ is the complexification of a real operator if and only if $\bar{A} = A$.

18.4. The exponential, determinant, and trace of a complex operator.

The exponential, determinant, and trace of a complex operator are defined in exactly the same way as in the real case, and they have exactly the same properties as in the real case, the only difference being that the determinant is now complex and hence not a volume.

Problem 1. Prove the following properties of the exponential:

$${}^{\mathbf{R}}(e^A) = e^{{}^{\mathbf{R}}A}, \qquad \overline{e^A} = e^{\bar{A}}, \qquad {}^{\mathbf{C}}(e^A) = e^{{}^{\mathbf{C}}A}.$$

Problem 2. Prove the following properties of the determinant:

$$\det {}^{\mathbf{R}}A = |\det A|^2, \qquad \det \bar{A} = \overline{\det A}, \qquad \det {}^{\mathbf{C}}A = \det A.$$

Problem 3. Prove the following properties of the trace:

$$\operatorname{Tr} {}^{\mathbf{R}}A = \operatorname{Tr} A + \operatorname{Tr} \bar{A}, \qquad \operatorname{Tr} \bar{A} = \overline{\operatorname{Tr} A}, \qquad \operatorname{Tr} {}^{\mathbf{C}}A = \operatorname{Tr} A.$$

Problem 4. Prove that the formula

$$\det e^A = e^{\operatorname{Tr} A}$$

continues to hold in the complex case.

18.5. The derivative of a curve with complex values.

By a *curve with complex values* is meant a mapping $\varphi\colon I \to \mathbf{C}^n$ of an open interval I of the real axis into the complex linear space $\mathbf{C}^n$. The *derivative* of the curve φ at a point $t_0 \in I$ is defined in the usual way and is a vector of the space $\mathbf{C}^n$:

$$\left.\frac{d\varphi}{dt}\right|_{t = t_0} = \lim_{h \to 0} \frac{\varphi(t_0 + h) - \varphi(t_0)}{h}.$$

Example 1. Let $n = 1$, $\varphi(t) = e^{it}$ (Fig. 112). Then

$$\left.\frac{d\varphi}{dt}\right|_{t = t_0} = i.$$

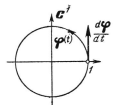

Fig. 112   The derivative of the curve $\varphi = e^{it}$ at the point 0 equals i.

Examining the case $n = 1$ in more detail, we note that curves with values in **C** can be multiplied as well as added, since multiplication is defined in **C**:

$$(\boldsymbol{\varphi}_1 + \boldsymbol{\varphi}_2)(t) = \boldsymbol{\varphi}_1(t) + \boldsymbol{\varphi}_2(t),$$
$$(\boldsymbol{\varphi}_1 \boldsymbol{\varphi}_2)(t) = \boldsymbol{\varphi}_1(t) \boldsymbol{\varphi}_2(t), \quad t \in I.$$

Problem 1. Prove that

$$\frac{d}{dt}(\boldsymbol{\varphi}_1 + \boldsymbol{\varphi}_2) = \frac{d\boldsymbol{\varphi}_1}{dt} + \frac{d\boldsymbol{\varphi}_2}{dt},$$

$$\frac{d}{dt}(\boldsymbol{\varphi}_1 \boldsymbol{\varphi}_2) = \frac{d\boldsymbol{\varphi}_1}{dt} \boldsymbol{\varphi}_2 + \boldsymbol{\varphi}_1 \frac{d\boldsymbol{\varphi}_2}{dt}.$$

Comment. In particular, the derivative of a polynomial with complex coefficients is given by the same formula as in the case of real coefficients.

If $n > 1$, we cannot multiply two curves with values in $\mathbf{C}^n$. However, since $\mathbf{C}^n$ is a **C**-module, we can multiply the curve $\boldsymbol{\varphi} : I \to \mathbf{C}^n$ by a function $f : I \to \mathbf{C}$:

$$(f\boldsymbol{\varphi})(t) = f(t)\boldsymbol{\varphi}(t).$$

Problem 2. Prove that

$$\frac{d}{dt}(^{\mathbf{R}}\boldsymbol{\varphi}) = \frac{^{\mathbf{R}}d\boldsymbol{\varphi}}{dt}, \qquad \frac{d}{dt}(^{\mathbf{C}}\boldsymbol{\varphi}) = \frac{^{\mathbf{C}}d\boldsymbol{\varphi}}{dt}, \qquad \frac{d\overline{\boldsymbol{\varphi}}}{dt} = \overline{\frac{d\boldsymbol{\varphi}}{dt}},$$

$$\frac{d(\boldsymbol{\varphi}_1 + \boldsymbol{\varphi}_2)}{dt} = \frac{d\boldsymbol{\varphi}_1}{dt} + \frac{d\boldsymbol{\varphi}_2}{dt}, \qquad \frac{d(f\boldsymbol{\varphi})}{dt} = \frac{df}{dt}\boldsymbol{\varphi} + f\frac{d\boldsymbol{\varphi}}{dt},$$

where, naturally, it is assumed that the derivatives in question exist.

THEOREM. *Let $A: \mathbf{C}^n \to \mathbf{C}^n$ be a **C**-linear operator. Then the **C**-linear operator*

$$\frac{d}{dt}e^{tA} = Ae^{tA}$$

from $\mathbf{C}^n$ into $\mathbf{C}^n$ exists for every $t \in \mathbf{R}$.

Proof. This can be proved in exactly the same way as in the real case, but we can also start from the real case. In fact, decomplexifying $\mathbf{C}^n$, we get

$$^{\mathbf{R}}\left(\frac{d}{dt}e^{tA}\right) = \frac{d}{dt}{}^{\mathbf{R}}(e^{tA}) = \frac{d}{dt}e^{t({}^{\mathbf{R}}A)} = ({}^{\mathbf{R}}A)e^{t({}^{\mathbf{R}}A)} = {}^{\mathbf{R}}(Ae^{tA}). \quad \blacksquare$$

19. Linear Equations with a Complex Phase Space

As often happens, the complex case is simpler than the real case. The complex case is important in its own right; moreover, investigation of the complex case will help us in our study of the real case.

19.1. Definitions. Let $A: \mathbf{C}^n \to \mathbf{C}^n$ be a **C**-linear operator. By a linear equation with phase space $\mathbf{C}^n$ is meant an equation

$$\dot{z} = Az, \qquad z \in \mathbf{C}^n. \tag{1}$$

The full description of (1) is "a system of homogeneous linear differential equations of the first order with constant complex coefficients."

By a *solution* $\boldsymbol{\varphi}$ of equation (1) satisfying the initial condition $\boldsymbol{\varphi}(t_0) = \mathbf{z}_0$, $t_0 \in \mathbf{R}$, $\mathbf{z}_0 \in \mathbf{C}^n$ is meant a mapping $\boldsymbol{\varphi}: I \to \mathbf{C}^n$ of an interval of the real t-axis into $\mathbf{C}^n$ such that $t_0 \in I$, $\boldsymbol{\varphi}(t_0) = \mathbf{z}_0$ and

$$\left.\frac{d\boldsymbol{\varphi}}{dt}\right|_{t=\tau} = A\boldsymbol{\varphi}(\tau)$$

for every $\tau \in I$. In other words, a mapping $\boldsymbol{\varphi}: I \to \mathbf{C}^n$ is said to be a solution of (1) if after decomplexifying the space $\mathbf{C}^n$ and the operator A, the mapping $\boldsymbol{\varphi}$ is a solution of the following equation with a $2n$-dimensional real phase space:

$$\dot{z} = {}^{\mathbf{R}}A\mathbf{z}, \qquad \mathbf{z} \in \mathbf{R}^{2n} = {}^{\mathbf{R}}\mathbf{C}^n.$$

19.2. The basic theorem. The following theorem is proved in exactly the same way as in the real case (see Theorems 15.2 and 15.3):

THEOREM. *The solution $\boldsymbol{\varphi}$ of equation (1) satisfying the initial condition $\boldsymbol{\varphi}(0) = \mathbf{z}_0$ is given by the formula $\boldsymbol{\varphi}(t) = e^{tA}\mathbf{z}_0$. Moreover, every one-parameter group $\{g^t, t \in \mathbf{R}\}$ of **C**-linear transformations of the space $\mathbf{C}^n$ is of the form*

$$g^t = e^{At}$$

*where $A: \mathbf{C}^n \to \mathbf{C}^n$ is a **C**-linear operator.*

Our goal is now to investigate and explicitly calculate e^{tA}.

19.3. The diagonal case. Let $A: \mathbf{C}^n \to \mathbf{C}^n$ be a $\mathbf{C}$-linear operator, and consider the characteristic equation

$$\det (A - \lambda E) = 0. \tag{2}$$

THEOREM. *If the n roots $\lambda_1, \ldots, \lambda_n$ of equation (2) are distinct, then $\mathbf{C}^n$ decomposes into a direct sum $\mathbf{C}^n = \mathbf{C}_1^1 + \cdots + \mathbf{C}_n^1$ of one-dimensional subspaces $\mathbf{C}_1^1, \ldots, \mathbf{C}_n^1$ invariant under A and e^{tA}, where in each one-dimensional invariant subspace, say $\mathbf{C}_k^1$, e^{tA} reduces to multiplication by the complex number $e^{\lambda_k t}$.*

Proof. The operator A has n linearly independent eigenlines:†

$$\mathbf{C}^n = \mathbf{C}_1^1 + \cdots + \mathbf{C}_n^1.$$

The operator A acts like multiplication by λ_k on the line $\mathbf{C}_k^1$, and hence the operator e^{tA} acts like multiplication by $e^{\lambda_k t}$. ∎

We now consider the one-dimensional case ($n = 1$) in more detail.

19.4. Example. Consider the linear equation

$$\dot{z} = \lambda z, \qquad z \in \mathbf{C}, \quad \lambda \in \mathbf{C}, \quad t \in \mathbf{R}, \tag{3}$$

with the complex line as its phase space. As we already know, the solution of (3) is just

$$\varphi(t) = e^{\lambda t} z_0.$$

Consider the complex function $e^{\lambda t}: \mathbf{R} \to \mathbf{C}$ of a real variable t. If λ is real, the function $e^{\lambda t}$ is real (Fig. 113), and the phase flow of equation (3) consists of expansion by a factor of $e^{\lambda t}$. If λ is purely imaginary, so that $\lambda = i\omega$, $\omega \in \mathbf{R}$, then by Euler's formula

$$e^{\lambda t} = e^{i\omega t} = \cos \omega t + i \sin \omega t.$$

In this case, the phase flow of equation (3) is a family $\{g^t\}$ of rotations through the angle ωt (Fig. 114). Finally, in the general case, $\lambda = \alpha + i\omega$

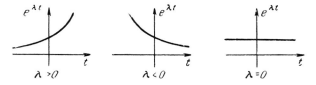

Fig. 113   Graph of the function $e^{\lambda t}$ for real t.

---

† This is the only place where the complex case differs from the real case. The greater complexity of the real case is due to the fact that the field $\mathbf{R}$ is not algebraically closed.

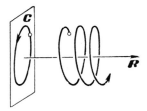

Fig. 114   Phase and integral curves of the equation $\dot{\mathbf{z}} = \lambda\mathbf{z}$ for purely imaginary λ.

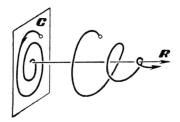

Fig. 115   Phase and integral curves of the equation $\dot{\mathbf{z}} = \lambda\mathbf{z}$ for $\lambda = \alpha + i\omega$, $\alpha < 0$, $\omega > 0$.

and multiplication by $e^{\lambda t}$ is the product of multiplication by e^{α} and multiplication by $e^{i\omega t}$ (see Sec. 15.5):

$$e^{\lambda t} = e^{(\alpha + i\omega)t} = e^{\alpha t}\, e^{i\omega t}. \tag{4}$$

The transformation g^t of the phase flow of equation (3) is then an $e^{\alpha t}$-fold expansion together with a simultaneous rotation through the angle ωt.

We now consider the phase curves in the general case. For example, suppose $\alpha < 0$, $\omega > 0$ (Fig. 115). Then as t increases, the phase point $e^{\lambda t}\mathbf{z}_0$ approaches the origin, winding around the origin "in the counterclockwise direction," i.e., from 1 to i. In polar coordinates, with a suitable choice of initial angle, the phase curve has equation

$$r = e^{k\theta}, \qquad k = \alpha/\omega$$

or

$$\theta = \frac{1}{k}\ln r.$$

A curve of this kind is called a *logarithmic spiral*. The phase curves are also logarithmic spirals for other combinations of the signs of α and ω (Figs. 116, 117). In every case (except $\lambda = 0$), the point $z = 0$ is the unique fixed point of the phase flow (and the unique singular point of the corresponding equa-

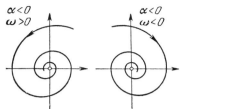

Fig. 116   A stable focus.   Fig. 117   An unstable focus.

Fig. 118   A center.

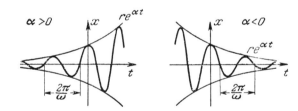

Fig. 119   The real part of $e^{\lambda t}$ as a function of time.

tion (3) of the vector field). This singular point is called a *focus* (we assume that $\alpha \neq 0$, $\omega \neq 0$). If $\alpha < 0$, then $\boldsymbol{\varphi}(t) \to 0$ as $t \to +\infty$, and the focus is said to be *stable*, while if $\alpha > 0$, the focus is said to be *unstable*. If $\alpha = 0$, $\omega \neq 0$, the phase curves are circles with the singular point as their *center* (Fig. 118).

Choosing the coordinate $z = x + iy$ in $\mathbf{C}^1$, we now investigate the change of the real and imaginary parts $x(t)$ and $y(t)$ as the phase point moves. It follows from (4) that

$$x(t) = re^{\alpha t} \cos(\omega t + \theta), \qquad y(t) = re^{\alpha t} \sin(\omega t + \theta),$$

where the constants r and θ are determined by the initial conditions (Fig. 119). Thus the coordinates $x(t)$ and $y(t)$ execute "harmonic oscillations of frequency ω with exponentially increasing amplitude $re^{\alpha t}$" if $\alpha > 0$, and damped oscillations if $\alpha < 0$. The change of x or y with time can also be written in the form

$$Ae^{\alpha t} \cos \omega t + Be^{\alpha t} \sin \omega t,$$

where the constants A and B are determined by the initial conditions.

Remark 1. By studying equation (3) in this way, we have simultaneously investigated all one-parameter groups of **C**-linear transformations of the complex line.

Remark 2. At the same time, we have investigated the system

$$\begin{cases} \dot{x} = \alpha x - \omega y, \\ \dot{y} = \omega x + \alpha y \end{cases}$$

of linear equations in the real plane obtained by decomplexifying equation (3).

Theorems 19.2 and 19.3, together with the above calculations, immediately imply an explicit formula for the solutions of equation (1).

19.5. Corollary. *Suppose the n roots $\lambda_1, \ldots, \lambda_n$ of the characteristic equation (2) are distinct. Then every solution φ of equation (1) is of the form*

$$\varphi(t) = \sum_{k=1}^{n} c_k e^{\lambda_k t} \xi_k, \tag{5}$$

where the ξ_k are constant vectors independent of the initial conditions and the c_k are complex constants depending on the initial conditions. For every choice of these constants, formula (5) gives a solution of equation (1).

Proof. We need only expand the initial condition with respect to the eigenbasis:

$$\varphi(0) = c_1 \xi_1 + \cdots + c_n \xi_n. \quad \blacksquare$$

If $z_1, \ldots, z_n$ is a linear system of coordinates in $\mathbf{C}_1^n$, then the real part x_l and the imaginary part y_l of every component of the solution $\varphi(t)$ changes with time like a linear combination of the functions $e^{\alpha_k t} \cos \omega_k t$ and $e^{\alpha_k t} \sin \omega_k t$, i.e.,

$$\begin{aligned} x_l &= \sum_{k=1}^{n} r_{kl} e^{\alpha_k t} (\cos \omega_k t + \theta_{kl}) \\ &= \sum_{k=1}^{n} A_{kl} e^{\alpha_k t} \cos \omega_k t + B_{kl} e^{\alpha_k t} \sin \omega_k t, \end{aligned} \tag{6}$$

where $\lambda_k = \alpha_k + i\omega_k$ and the various r, θ, A, B are real constants depending on the initial conditions.

20. Complexification of a Real Linear Equation

We now use the results of our study of the complex equation to investigate the real case.

20.1. The complexified equation. Let $A: \mathbf{R}^n \to \mathbf{R}^n$ be a linear operator, specifying a linear equation

$$\dot{\mathbf{x}} = A\mathbf{x}, \qquad \mathbf{x} \in \mathbf{R}^n. \tag{1}$$

The complexification of equation (1) is the equation

$$\dot{\mathbf{z}} = {}^{\mathbf{C}}A\mathbf{z}, \qquad \mathbf{z} \in \mathbf{C}^n = {}^{\mathbf{C}}\mathbf{R}^n \tag{2}$$

with a complex phase space.

LEMMA 1. *The solutions of equation (2) with complex conjugate initial conditions are themselves complex conjugates.*

Proof. If $\boldsymbol{\varphi}$ is the solution with initial condition $\boldsymbol{\varphi}(t_0) = \mathbf{z}_0$ (Fig. 120), then $\bar{\boldsymbol{\varphi}}(t_0) = \bar{\mathbf{z}}_0$. Once we show that $\bar{\boldsymbol{\varphi}}$ is a solution, the lemma will be proved (because of the uniqueness). But

$$\frac{d\bar{\boldsymbol{\varphi}}}{dt} = \overline{\frac{d\boldsymbol{\varphi}}{dt}} = \overline{{}^{\mathbf{C}}A\boldsymbol{\varphi}} = \overline{{}^{\mathbf{C}}A}\,\bar{\boldsymbol{\varphi}} = {}^{\mathbf{C}}A\,\bar{\boldsymbol{\varphi}}. \quad \blacksquare$$

Remark. Instead of equation (2), we might have chosen the more general equation

$$\dot{\mathbf{z}} = F(\mathbf{z}, t), \qquad \mathbf{z} \in {}^{\mathbf{C}}\mathbf{R}^n,$$

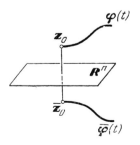

Fig. 120   Complex conjugate solutions.

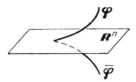

Fig. 121   A solution with a real initial condition cannot have complex values.

whose right-hand side takes complex conjugate values at complex conjugate points:

$$F(\bar{\mathbf{z}}, t) = \overline{F(\mathbf{z}, t)}.$$

For example, this condition is satisfied by any polynomial in the coordinates z_k of the vector $\mathbf{z}$ in a real basis whose coefficients are real functions of t.

COROLLARY. *The solution of equation* (2) *with a real initial condition is real and satisfies equation* (1).

Proof. If $\bar{\varphi} \neq \varphi$ (Fig. 121), the uniqueness theorem would be violated. ∎

In the next lemma, the linearity of the equation is essential.

LEMMA 2. *The function* $\mathbf{z} = \varphi(t)$ *is a solution of the complexified equation* (2) *if and only if its real and imaginary parts satisfy the original equation* (1).

Proof. Since

$$^{\mathbf{C}}A(\mathbf{x} + i\mathbf{y}) = A\mathbf{x} + iA\mathbf{y},$$

the decomplexified equation (2) decomposes into a direct product

$$\begin{cases} \dot{\mathbf{x}} = A\mathbf{x}, & \mathbf{x} \in \mathbf{R}^n, \\ \dot{\mathbf{y}} = A\mathbf{y}, & \mathbf{y} \in \mathbf{R}^n. \end{cases} \quad ∎$$

It is clear from Lemmas 1 and 2 that from a knowledge of the complex solutions of equation (2) we can find the real solutions of equation (1), and conversely. In particular, *formula* (6) *of Sec. 19.5 gives the explicit form of the solution in the case where the characteristic equation has no multiple roots.*

20.2. Invariant subspaces of a real operator. Let $A \colon \mathbf{R}^n \to \mathbf{R}^n$ be a real linear operator, and let λ be one of the roots (in general complex) of the characteristic equation $\det(A - \lambda E) = 0$.

LEMMA 3. *If* $\xi \in \mathbf{C}^n = {}^{\mathbf{C}}\mathbf{R}^n$ *is an eigenvector of the operator* ${}^{\mathbf{C}}A$ *with eigenvalue* λ, *then* $\bar{\xi}$ *is an eigenvector with eigenvalue* $\bar{\lambda}$. *Moreover,* λ *and* $\bar{\lambda}$ *have the same multiplicity.*

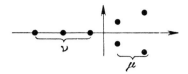

Fig. 122   Eigenvalues of a real operator.

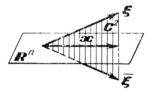

Fig. 123   The real part of an eigenvector belongs to an invariant real plane.

Proof. Since $\overline{{}^{\mathbf{C}}A} = {}^{\mathbf{C}}A$, the equation ${}^{\mathbf{C}}A\xi = \lambda\xi$ is equivalent to ${}^{\mathbf{C}}A\bar{\xi} = \bar{\lambda}\bar{\xi}$ and the characteristic equation has real coefficients. ∎

Suppose now that the eigenvalues $\lambda_1, \ldots, \lambda_n \in \mathbf{C}$ of the operator $A : \mathbf{R}^n \to \mathbf{R}^n$ are distinct (Fig. 122). Among these eigenvalues we have a certain number ν of real eigenvalues and a certain number μ of complex conjugate pairs (where $\nu + 2\mu = n$, so that the parity of the number of real eigenvalues equals the parity of n).

THEOREM. *The space $\mathbf{R}^n$ decomposes into a direct sum of ν one-dimensional subspaces invariant under A and μ two-dimensional subspaces invariant under A.*

Proof. To every real eigenvalue there corresponds a real eigenvector and hence a one-dimensional invariant subspace in $\mathbf{R}^n$. Let $\lambda, \bar{\lambda}$ be a pair of complex conjugate eigenvectors. Then to λ there corresponds an eigenvector $\xi \in \mathbf{C}^n = {}^{\mathbf{C}}\mathbf{R}^n$ of the complexified operator ${}^{\mathbf{C}}A$. By Lemma 3, the complex conjugate vector $\bar{\xi}$ is also an eigenvector, with eigenvalue $\bar{\lambda}$. The complex plane $\mathbf{C}^2$ spanned by the eigenvectors $\xi, \bar{\xi}$ is invariant under the operator ${}^{\mathbf{C}}A$, and the real subspace $\mathbf{R}^n \subset {}^{\mathbf{C}}\mathbf{R}^n$ is also invariant under ${}^{\mathbf{C}}A$. Hence their intersection is also invariant under ${}^{\mathbf{C}}A$. We now show that this intersection is a two-dimensional real plane $\mathbf{R}^2$ (Fig. 123).

To this end, consider the real and imaginary parts of the eigenvector ξ:

$$\mathbf{x} = \frac{1}{2}(\xi + \bar{\xi}) \in \mathbf{R}^n, \qquad \mathbf{y} = \frac{1}{2i}(\xi - \bar{\xi}) \in \mathbf{R}^n.$$

Being $\mathbf{C}$-linear combinations of the vectors ξ and $\bar{\xi}$, the vectors $\mathbf{x}$ and $\mathbf{y}$ belong to the intersection $\mathbf{C}^2 \cap \mathbf{R}^n$. The vectors $\mathbf{x}$ and $\mathbf{y}$ are $\mathbf{C}$-linearly independent, since the **C**-independent vectors ξ and $\bar{\xi}$ are linear combinations of **x** and **y**:

$$\xi = \mathbf{x} + i\mathbf{y}, \qquad \bar{\xi} = \mathbf{x} - i\mathbf{y}.$$

Hence every vector η of the plane $\mathbf{C}^2$ has a unique representation as a complex linear combination of the real vectors **x** and **y**:

$$\eta = a\mathbf{x} + b\mathbf{y}, \qquad a, b \in \mathbf{C}.$$

Such a vector is real ($\eta = \bar{\eta}$) if and only if $\bar{a}\mathbf{x} + \bar{b}\mathbf{y} = a\mathbf{x} + b\mathbf{y}$, i.e., if and only if a and b are real. Thus *the intersection* $\mathbf{C}^2 \cap \mathbf{R}^n$ *is the two-dimensional real plane spanned by the vectors* **x** *and* **y** *which are the real and imaginary parts of the eigenvector* ξ. Moreover, λ and $\bar{\lambda}$ are the eigenvalues of the restriction of the operator A to the plane $\mathbf{R}^2$. In fact, complexification does not change eigenvalues. After complexifying the restriction of A to $\mathbf{R}^2$, we get the restriction of $^{\mathbf{C}}A$ to $\mathbf{C}^2$. But the plane $\mathbf{C}^2$ is spanned by the eigenvectors of the operator $^{\mathbf{C}}A$ with eigenvalues λ and $\bar{\lambda}$. Hence λ and $\bar{\lambda}$ are the eigenvalues of the restriction of A to $\mathbf{R}^2$.

We must still show that the one-dimensional and two-dimensional subspaces of $\mathbf{R}^n$ just constructed are **R**-linearly independent. But this follows at once from the fact that the n eigenvectors of the operator $^{\mathbf{C}}A$ are **C**-linearly independent and can be expressed linearly in terms of the vectors ξ_k ($k = 1, \ldots, \nu$) and $\mathbf{x}_k, \mathbf{y}_k$ ($k = 1, \ldots, \mu$). ∎

Thus *in the case where all the eigenvalues of the operator* $A: \mathbf{R}^n \to \mathbf{R}^n$ *are simple, the linear differential equation*

$$\dot{\mathbf{x}} = A\mathbf{x}, \qquad \mathbf{x} \in \mathbf{R}^n$$

decomposes into a direct product of equations with one-dimensional and two-dimensional phase spaces.

We note that a polynomial with "general" coefficients has no multiple roots. Hence, to investigate linear differential equations, we must first of all consider linear differential equations on the line (as we have already done) and in the plane.

20.3. Linear equations in the plane.

THEOREM. *Let* $A: \mathbf{R}^2 \to \mathbf{R}^2$ *be a linear operator with complex eigenvalues* λ, $\bar{\lambda}$. *Then* A *is the decomplexification of the operator* $\Lambda: \mathbf{C}^1 \to \mathbf{C}^1$ *of multiplication by the complex number* λ. *More exactly, the plane* $\mathbf{R}^2$ *can be equipped with the structure of the line* $\mathbf{C}^1$, *so that* $\mathbf{R}^2 = {}^{\mathbf{R}}\mathbf{C}^1$ *and* $A = {}^{\mathbf{R}}\Lambda$.

Proof. The proof consists of a rather mysterious calculation.† Let $\mathbf{x} + i\mathbf{y} \in {}^{\mathbf{C}}\mathbf{R}^2$ be a complex eigenvector of the operator ${}^{\mathbf{C}}A$ with eigenvalue $\lambda = \alpha + i\omega$. The vectors $\mathbf{x}$ and $\mathbf{y}$ form a basis in $\mathbf{R}^2$. On the one hand, we have

$${}^{\mathbf{C}}A(\mathbf{x} + i\mathbf{y}) = (\alpha + i\omega)(\mathbf{x} + i\mathbf{y}) = \alpha\mathbf{x} - \omega\mathbf{y} + i(\omega\mathbf{x} + \alpha\mathbf{y}),$$

while on the other,

$${}^{\mathbf{C}}A(\mathbf{x} + i\mathbf{y}) = A\mathbf{x} + iA\mathbf{y},$$

and hence

$$A\mathbf{x} = \alpha\mathbf{x} - \omega\mathbf{y}, \qquad A\mathbf{y} = \omega\mathbf{x} + \alpha\mathbf{y},$$

i.e., the operator $A: \mathbf{R}^2 \to \mathbf{R}^2$ in the basis $\mathbf{x}, \mathbf{y}$ has the same matrix

$$\begin{pmatrix} \alpha & \omega \\ -\omega & \alpha \end{pmatrix}$$

as the operator ${}^{\mathbf{R}}\Lambda$ of multiplication by $\lambda = \alpha + i\omega$ in the basis $1, -i$. Thus the desired complex structure in $\mathbf{R}^2$ is obtained by taking $\mathbf{x}$ for 1 and $\mathbf{y}$ for $-i$. ∎

COROLLARY 1. *Let $A: \mathbf{R}^2 \to \mathbf{R}^2$ be a linear transformation of the Euclidean plane with complex eigenvalues $\lambda, \bar{\lambda}$. Then the transformation A is affinely equivalent to a $|\lambda|$-fold expansion together with simultaneous rotation through the angle $\arg \lambda$.*

COROLLARY 2. *The phase flow of the linear equation* (1) *in the Euclidean plane $\mathbf{R}^2$ with complex eigenvalues $\lambda, \bar{\lambda} = \alpha \pm i\omega$ is affinely equivalent to a family of $e^{\alpha t}$-fold expansions with simultaneous rotation through the angle ωt.*

In particular, the singular point 0 is a focus, while the phase curves are affine images of logarithmic spirals approaching the origin as $t \to +\infty$ in the case where the real part α of the eigenvalues $\lambda, \bar{\lambda}$ is negative and moving away from the origin in the case where $\alpha > 0$ (Fig. 124).

In the case $\alpha = 0$ (Fig. 125), the phase curves are a family of concentric ellipses, with the singular point as their *center*. In this case, the transformations are called *elliptical rotations*.

20.4. Classification of singular points in the plane. Now let

$$\dot{\mathbf{x}} = A\mathbf{x}, \qquad \mathbf{x} \in \mathbf{R}^2, \quad A: \mathbf{R}^2 \to \mathbf{R}^2$$

† The calculation can be replaced by the following argument. Let $\lambda = \alpha + i\omega$, and define an operator $I: \mathbf{R}^2 \to \mathbf{R}^2$ by the condition $A = \alpha E + \omega I$. Such an operator I exists, since $\omega \neq 0$ by hypothesis. Then $I^2 = -E$, since the operator A satisfies its own characteristic equation. Taking I to be multiplication by i, we get the necessary complex structure in $\mathbf{R}^2$.

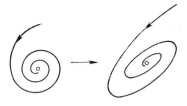

Fig. 124   The affine image of a
logarithmic spiral.

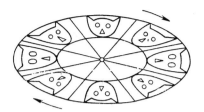

Fig. 125   An elliptical rotation.

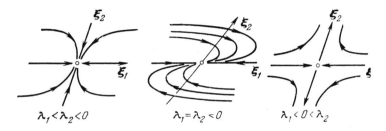

Fig. 126   An unstable focus.

Fig. 127   A saddle point.

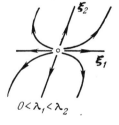

Fig. 128   An unstable node.

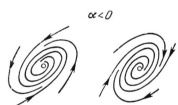

Fig. 129   Stable foci.

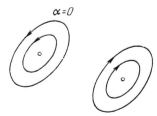

Fig. 130   Centers.

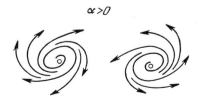

Fig. 131   Unstable foci.

be any linear equation in the plane, and suppose the roots λ_1, λ_2 of the characteristic equation are distinct. If the roots are real and $\lambda_1 < \lambda_2$, the equation decomposes into two one-dimensional equations and we get one of the cases already studied in Chap. 1 (Figs. 126, 127, 128).

Here we omit the boundary cases where λ_1 or λ_2 equals 0. These cases are of much less interest, since they are rarely encountered and are not preserved under an arbitrarily small perturbation; they can be investigated with no difficulty whatsoever.

If the roots are complex, so that $\lambda_{1,2} = \alpha \pm i\omega$, then, depending on the sign of α, we get one of the cases shown in Figs. 129, 130, 131. The case of a center is exceptional, but is encountered, for example, in conservative systems (see Sec. 12). The case of multiple roots is also exceptional. As an exercise, the reader should verify that the case shown in Fig. 126 corresponds to a Jordan block with $\lambda_1 = \lambda_2 < 0$ (a "degenerate node").

20.5. Example: The pendulum with friction. We now apply everything said to the equation

$$\ddot{x} = -x - k\dot{x}$$

of small oscillations of a pendulum with friction (k is the coefficient of friction). The equivalent system

$$\begin{cases} \dot{x}_1 = x_2, \\ \dot{x}_2 = -x_1 - kx_2, \end{cases}$$

has the matrix

$$\begin{pmatrix} 0 & 1 \\ -1 & -k \end{pmatrix},$$

with determinant 1 and trace $-k$. The corresponding characteristic equation

$$\lambda^2 + k\lambda + 1 = 0$$

has complex roots if $|k| < 2$, i.e., if the friction is not too large.†

The real part of each of the complex roots

$$\lambda_{1,2} = \alpha \pm i\omega$$

equals $-k/2$. Hence *if the coefficient of friction is positive and not too large* $(0 < k < 2)$, *the lower equilibrium position of the pendulum* $(x_1 = x_2 = 0)$ *is a stable focus.* As $k \to 0$, the focus becomes a center. The smaller the coefficient of friction, the slower the phase point approaches the equilibrium position

---

† The case of real roots is considered in Sec. 17.2.

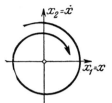

Fig. 132   Phase curve of a pendulum with small friction.

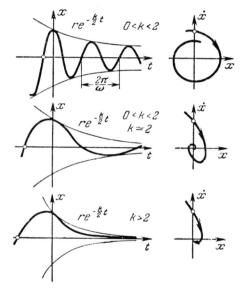

Fig. 133   The transition from damped oscillations to nonoscillatory motion of the pendulum: Phase curves and graphs of solutions for three values of the coefficient of friction.

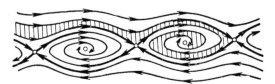

Fig. 134   Phase plane of the pendulum with small friction. After a certain number of revolutions, the pendulum begins to swing near the lower equilibrium position.

as $t \to +\infty$ (Fig. 132). Explicit formulas for the change of $x_1 = x$ with time can be obtained from Corollary 2 of Sec. 20.3 and the formulas of Sec. 19.4. Thus

$$x(t) = re^{\alpha t} \cos(\omega t - \theta) = Ae^{\alpha t} \cos \omega t + Be^{\alpha t} \sin \omega t,$$

where the coefficients r and θ (or A and B) can be determined from the initial conditions.

Hence the oscillations of the pendulum are damped, with variable amplitude $re^{\alpha t}$ and period $2\pi/\omega$. The larger the coefficient of friction, the faster the amplitude decreases.† The frequency

$$\omega = \sqrt{1 - \frac{k^2}{4}}$$

decreases as the coefficient of friction k increases. As $k \to 2$, the frequency approaches 0 and the period approaches ∞ (Fig. 133). For small k, we have

$$\omega \approx 1 - \frac{k^2}{8}, \qquad k \to 0.$$

Thus the friction increases the period only very slightly, and its influence on the frequency can be neglected in many calculations.

Problem 1. Draw phase curves for the nonlinearized pendulum with friction

$$\ddot{x} = -\sin x - k\dot{x}$$

(Fig. 134).

Hint. Calculate the derivative of the total energy along the phase curve.

20.6. General solution of the linear equation in the case of simple roots of the characteristic equation. We already know that every solution φ of the complexified equation is a linear combination

$$\varphi(t) = \sum_{k=1}^{n} c_k e^{\lambda_k t} \xi_k$$

of exponentials (see Sec. 19.5), where ξ_k is any eigenvector with eigenvalue λ_k; *here we choose real eigenvectors if the eigenvalues are real and complex conjugate eigenvectors if the eigenvalues are complex conjugates.* Moreover, we also know that the solutions of the real equation are the solutions of its complexification with real initial conditions. A necessary and sufficient condition for the vector $\varphi(0)$ to be real is that

$$\sum_{k=1}^{n} c_k \xi_k = \sum_{k=1}^{n} \bar{c}_k \bar{\xi}_k.$$

† Nevertheless, the pendulum still makes infinitely many swings for any value $k < 2$. If $k > 2$, however, the pendulum changes its direction of motion no more than once.

For this to hold, *the coefficients of real vectors must be real and those of complex conjugate vectors must be complex conjugates.* Note further that the n complex constants $c_1, \ldots, c_n$ are uniquely determined by the solution of the complex equation (for a fixed choice of eigenvectors). This proves the following

THEOREM. *Every solution of the real equation has a unique representation of the form*

$$\boldsymbol{\varphi}(t) = \sum_{k=1}^{\nu} a_k e^{\lambda_k t} \bar{\boldsymbol{\xi}}_k + \sum_{k=\nu+1}^{\nu+\mu} (c_k e^{\lambda_k t} \boldsymbol{\xi}_k + \bar{c}_k e^{\bar{\lambda}_k t} \bar{\boldsymbol{\xi}}_k) \tag{3}$$

(for a fixed choice of eigenvectors), where the a_k are real constants and the c_k complex constants.

Formula (3) is called the *general solution* of the equation. We can also write (3) in the form

$$\boldsymbol{\varphi}(t) = \sum_{k=1}^{\nu} a_k e^{\lambda_k t} \boldsymbol{\xi}_k + 2 \operatorname{Re} \sum_{k=\nu+1}^{\nu+\mu} c_k e^{\lambda_k t} \boldsymbol{\xi}_k.$$

Note that the general solution depends on $\nu + 2\mu = n$ real constants a_k, $\operatorname{Re} c_k$, and $\operatorname{Im} c_k$. These constants are uniquely determined by the initial conditions.

COROLLARY 1. *Let $\boldsymbol{\varphi} = (\varphi_1, \ldots, \varphi_n)$ be a solution of a system of n real linear differential equations of the first order with matrix A, and suppose all the roots of the characteristic equation of the matrix A are simple, where the real roots are denoted by λ_k and the complex roots by $\alpha_k \pm i\omega_k$. Then every function φ_m is a linear combination of the functions*

$$e^{\lambda_k t}, \qquad e^{\alpha_k t} \cos \omega_k t, \qquad e^{\alpha_k t} \sin \omega_k t. \tag{4}$$

Proof. Let $\boldsymbol{\varphi} = \xi_1 \mathbf{e}_1 + \cdots + \xi_n \mathbf{e}_n$ be the expansion of the general solution (3) with respect to the coordinate basis $\mathbf{e}_1, \ldots, \mathbf{e}_n$, bearing in mind that

$$e^{(\alpha_k \pm i\omega_k)t} = e^{\alpha_k t}(\cos \omega_k t \pm i \sin \omega_k t). \quad \blacksquare$$

To solve linear systems in practice, we can use the method of undetermined coefficients to look for solutions in the form of linear combinations of the functions (4).

COROLLARY 2. *Let A be a real matrix with real eigenvalues λ_k and complex eigenvalues $\alpha_k \pm i\omega_k$, all of which are simple. Then every element of the matrix e^{tA} is a linear combination of the functions (4).*

Proof. Every column of the matrix e^{tA} is made up of components of the image of a basis vector under the action of the phase flow of the system of differential equations with matrix A. $\blacksquare$

Remark. Everything said above immediately carries over to equations and systems of equations of order higher than 1, inasmuch as these reduce to systems of first-order equations (see Sec. 9).

Problem 1. Find all real solutions of the equations

$$x^{(iv)} + 4x = 0, \qquad x^{(iv)} = x, \qquad \ddot{x} + x = 0.$$

21. Classification of Singular Points of Linear Systems

As shown above, the general real linear system (whose characteristic equation has no multiple roots) reduces to a direct product of one-dimensional and two-dimensional systems. Since one-dimensional and two-dimensional systems have already been studied, we are now in a position to investigate multidimensional systems.

21.1. Example: Singular points in three-dimensional space. Here the characteristic equation is a real cubic equation. Such an equation can have either three real roots or one real and two complex roots, and many different cases can occur, depending on the arrangement of the roots $\lambda_1, \lambda_2, \lambda_3$ in the plane of the complex variable λ. Examining the order and signs of the real parts of the roots, we find that 10 "nondegenerate" cases (Fig. 135) are possible, as well as a number of "degenerate" cases (see e.g., Fig. 136),

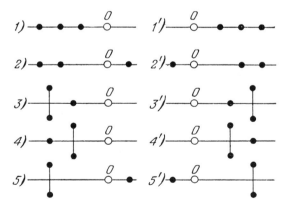

Fig. 135 Eigenvalues of a real operator $A: \mathbf{R}^3 \to \mathbf{R}^3$. Nondegenerate cases.

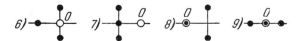

Fig. 136 Some degenerate cases.

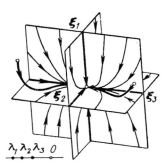

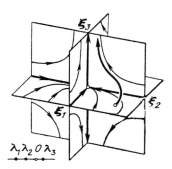

Fig. 137 Phase space of a linear equation in the case $\lambda_1 < \lambda_2 < \lambda_3 < 0$. The phase flow is a contraction in all three directions.

Fig. 138 The case $\lambda_1 < \lambda_2 < 0 < \lambda_3$: Contraction in two directions and expansion in the third.

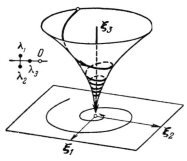

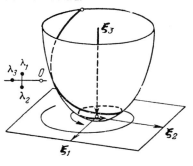

Fig. 139 The case $\operatorname{Re}\lambda_{1,2} < \lambda_3 < 0$: Contraction in the direction of $\boldsymbol{\xi}_3$ and rotation with faster contraction in the plane of $\boldsymbol{\xi}_1$ and $\boldsymbol{\xi}_2$.

Fig. 140 The case $\lambda_3 < \operatorname{Re}\lambda_{1,2} < 0$: Contraction in the direction of $\boldsymbol{\xi}_3$ and rotation with slower contraction in the plane of $\boldsymbol{\xi}_1$ and $\boldsymbol{\xi}_2$.

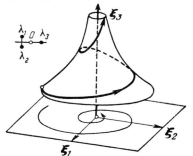

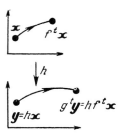

Fig. 141 The case $\operatorname{Re}\lambda_{1,2} < 0 < \lambda_3$: Expansion in the direction of $\boldsymbol{\xi}_3$ and rotation with contraction in the plane of $\boldsymbol{\xi}_1$ and $\boldsymbol{\xi}_2$.

Fig. 142 Equivalent flows.

where the real part of one of the roots λ_k vanishes or equals the real part of a root not conjugate to λ_k (here we do not consider the case of multiple roots). The investigation of the behavior of the phase curves in each of these cases presents no difficulty.

Bearing in mind that $e^{\lambda t}$ approaches 0 as $t \to +\infty$ if $\operatorname{Re} \lambda < 0$ (the more rapidly, the smaller $\operatorname{Re} \lambda$), we get the phase curves

$$\boldsymbol{\varphi}(t) = \operatorname{Re}(c_1 e^{\lambda_1 t}\boldsymbol{\xi}_1 + c_2 e^{\lambda_2 t}\boldsymbol{\xi}_2 + c_3 e^{\lambda_3 t}\boldsymbol{\xi}_3)$$

shown in Figs. 137–141. Cases 1')–5') are obtained from cases 1)–5) by changing the direction of the t-axis, so that correspondingly we need only reverse the direction of all the arrows in Figs. 137–141.

Problem 1. Draw phase curves for cases 6)–9) in Fig. 136.

21.2. Linear, differentiable, and topological equivalence. Each of
these classifications is based on some equivalence relation. There exist at least three reasonable equivalence relations for linear systems, corresponding to algebraic, differentiable, and topological mappings.

Definition. Two phase flows $\{f^t\}, \{g^t\}: \mathbf{R}^n \to \mathbf{R}^n$ are said to be *equivalent†* if there exists a one-to-one mapping $h: \mathbf{R}^n \to \mathbf{R}^n$ carrying the flow $\{f^t\}$ into the flow $\{g^t\}$ such that $h \circ f^t = g^t \circ h$ for every $t \in \mathbf{R}$ (Fig. 142). (We then say that "the flow $\{f^t\}$ is transformed into the flow $\{g^t\}$ by the change of coordinates h.") Under these conditions, the flows are said to be

1) *Linearly equivalent* if the mapping $h: \mathbf{R}^n \to \mathbf{R}^n$ in question is a *linear automorphism*;

2) *Differentiably equivalent* if the mapping $h: \mathbf{R}^n \to \mathbf{R}^n$ is a *diffeomorphism*;

3) *Topologically equivalent* if the mapping $h: \mathbf{R}^n \to \mathbf{R}^n$ is a *homeomorphism*, i.e., if h is one-to-one and continuous in both directions.

Problem 1. Prove that linear equivalence implies differentiable equivalence, while differentiable equivalence implies topological equivalence.

Remark. Note that the mapping h carries phase curves of the flow $\{f^t\}$ into phase curves of the flow $\{g^t\}$.

Problem 2. Does every linear automorphism $h \in GL(\mathbf{R}^n)$ carrying phase curves of the flow $\{f^t\}$ into phase curves of the flow $\{g^t\}$ establish a linear equivalence between the flows?

Ans. No.

Hint. Let $n = 1, f^t x = e^t x, g^t x = e^{2t} x.$

† The terms "conjugate" and "similar" are sometimes used as synonyms for "equivalent" as defined here.

Problem 3. Prove that the relations of linear, differentiable, and topological equivalence are actually equivalence relations, i.e., that

$$f \sim f, \qquad f \sim g \Rightarrow g \sim f, \qquad f \sim g, g \sim k \Rightarrow f \sim k.$$

In particular, everything said is applicable to the phase flows of linear systems. For brevity, we will talk about equivalence of the systems themselves. Thus we have divided all linear systems into equivalence classes in three ways, corresponding to linear, differentiable, and topological equivalence. We now study these classes in more detail.

21.3. The linear classification.

THEOREM. *Let A, $B: \mathbf{R}^n \to \mathbf{R}^n$ be linear operators all of whose eigenvalues are simple. Then the systems*

$$\dot{\mathbf{x}} = A\mathbf{x}, \qquad \mathbf{x} \in \mathbf{R}^n,$$
$$\dot{\mathbf{y}} = B\mathbf{y}, \qquad \mathbf{y} \in \mathbf{R}^n$$

are linearly equivalent if and only if the eigenvalues of the operators A and B coincide.

Proof. A necessary and sufficient condition for linear equivalence of linear systems is that $B = hAh^{-1}$ for some $h \in GL(\mathbf{R}^n)$, since $\dot{\mathbf{y}} = h\dot{\mathbf{x}} = hA\mathbf{x} = hAh^{-1}\mathbf{y}$ (Fig. 143). But the eigenvalues of the operators A and hAh^{-1} coincide (here simplicity of the eigenvalues is unimportant).

Conversely, suppose the eigenvalues of A are simple and coincide with the eigenvalues of B. Then, according to Sec. 20, A and B decompose into direct products of identical (linearly equivalent) one-dimensional and two-dimensional systems. Therefore A and B are linearly equivalent. ∎

Problem 1. Show that the systems

$$\begin{cases} \dot{x}_1 = x_1, \\ \dot{x}_2 = x_2, \end{cases} \qquad \begin{cases} \dot{x}_1 = x_1 + x_2, \\ \dot{x}_2 = x_2 \end{cases}$$

are not linearly equivalent, even though their eigenvalues coincide.

21.4. The differentiable classification. Our next theorem is almost obvious:

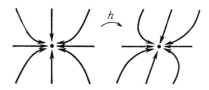

Fig. 143   Linearly equivalent systems.

THEOREM. *Two linear systems*

$$\dot{\mathbf{x}} = A\mathbf{x}, \qquad \dot{\mathbf{x}} = B\mathbf{x}, \qquad \mathbf{x} \in \mathbf{R}^n$$

are differentiably equivalent if and only if they are linearly equivalent.†

Proof. Let $h: \mathbf{R}^n \to \mathbf{R}^n$ be a diffeomorphism carrying the phase flow of the system A into the phase flow of the system B. The point $\mathbf{x} = 0$ is a fixed point of the phase flow of the system A. Therefore h carries 0 into one of the fixed points $\mathbf{c}$ of the phase flow of the system B, so that $B\mathbf{c} = 0$. The diffeomorphism $\delta: \mathbf{R}^n \to \mathbf{R}^n$ of shift by $\mathbf{c}$ ($\delta\mathbf{x} = \mathbf{x} - \mathbf{c}$) carries the phase flow of B into itself, since

$$\frac{d}{dt}(\mathbf{x} - \mathbf{c}) = \dot{\mathbf{x}} = B\mathbf{x} = B(\mathbf{x} - \mathbf{c}),$$

while the diffeomorphism

$$h_1 = \delta \circ h : \mathbf{R}^n \to \mathbf{R}^n$$

carries the flow of A into the flow of B and leaves 0 fixed: $h_1(0) = 0$.

Now let $H: \mathbf{R}^n \to \mathbf{R}^n$ be the derivative of the diffeomorphism h_1 at 0, so that $H = h_{*}|_0 \in GL(\mathbf{R}^n)$. The diffeomorphisms $h_1 \circ e^{tA}$ and $e^{tB} \circ h_1$ coincide for every t, and hence so do their derivatives at $\mathbf{x} = 0$:

$$He^{tA} = e^{tB}H. \quad \blacksquare$$

22. Topological Classification of Singular Points

Consider two linear systems

$$\dot{\mathbf{x}} = A\mathbf{x}, \qquad \dot{\mathbf{x}} = B\mathbf{x}, \qquad \mathbf{x} \in \mathbf{R}^n,$$

all of whose eigenvalues have nonzero real parts. Let m_- denote the number of eigenvalues with a negative real part and m_+ the number with a positive real part, so that $m_- + m_+ = n$.

22.1. Theorem. *A necessary and sufficient condition for topological equivalence of two linear systems, all of whose eigenvalues have nonzero real parts, is that the number of eigenvalues with negative (and hence positive) real parts be the same in both systems:*

$$m_-(A) = m_-(B), \qquad m_+(A) = m_+(B).$$

† It must not be thought that every diffeomorphism establishing the equivalence of the systems is linear (for example, let $A = B = 0$).

Fig. 144   Topologically equivalent and nonequivalent systems.

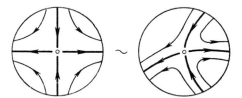

Fig. 145   Topological equivalence of a system and of its linearization.

For example, this theorem asserts that stable nodes and foci (Fig. 144) are topologically equivalent to each other $(m_- = 2)$ but not topologically equivalent to a saddle point $(m_- = m_+ = 2)$.

Just like the index of inertia of a nondegenerate quadratic form, the number m_- (or m_+) is the unique topological invariant of a linear system.

Remark. A similar result holds *locally* (in a neighborhood of a fixed point) for *nonlinear* systems whose linear parts have no purely imaginary eigenvalues. In particular, in a neighborhood of a fixed point such a system is topologically equivalent to its linear part (Fig. 145). Here we will not go into the proof of this proposition, of great importance in the study of nonlinear systems.

22.2. Reduction to the case $m_- = 0$. The topological equivalence of linear systems with identical values of m_- and m_+ is a consequence of the following three lemmas:

LEMMA 1. *Direct products of topologically equivalent systems are topologically equivalent. More exactly, if the systems specified by the operators*

$$A_1, B_1 : \mathbf{R}^{m_1} \to \mathbf{R}^{m_1}, \qquad A_2, B_2 : \mathbf{R}^{m_2} \to \mathbf{R}^{m_2}$$

are carried into each other by the homeomorphisms

$$h_1 : \mathbf{R}^{m_1} \to \mathbf{R}^{m_1}, \qquad h_2 : \mathbf{R}^{m_2} \to \mathbf{R}^{m_2},$$

then there exists a homeomorphism

$$h : \mathbf{R}^{m_1} \dotplus \mathbf{R}^{m_2} \to \mathbf{R}^{m_1} \dotplus \mathbf{R}^{m_2}$$

carrying the phase flow of the product system

$$\dot{\mathbf{x}}_1 = A_1\mathbf{x}_1, \qquad \dot{\mathbf{x}}_2 = A_2\mathbf{x}_2$$

into the phase flow of the product system

$$\dot{\mathbf{x}}_1 = B_1\mathbf{x}_1, \qquad \dot{\mathbf{x}}_2 = B_2\mathbf{x}_2.$$

Proof. Simply let

$$h(\mathbf{x}_1, \mathbf{x}_2) = (h_1(\mathbf{x}_1), h_2(\mathbf{x}_2)). \quad \blacksquare$$

The next lemma is familiar from a course in linear algebra.

LEMMA 2. *If the operator $A: \mathbf{R}^n \to \mathbf{R}^n$ has no purely imaginary eigenvalues, then the space $\mathbf{R}^n$ decomposes into the direct sum $\mathbf{R}^n = \mathbf{R}^{m-} \dotplus \mathbf{R}^{m+}$ of two subspaces invariant under the operator A, such that all the eigenvalues of the restriction of A to $\mathbf{R}^{m-}$ have negative real parts and all the eigenvalues of the restriction of A to $\mathbf{R}^{m+}$ have positive real parts (Fig. 146).*

Proof. This follows, for example, from the theorem on the Jordan normal form. $\blacksquare$

Lemmas 1 and 2 reduce the proof of topological equivalence to the following special case:

LEMMA 3. *Let $A: \mathbf{R}^n \to \mathbf{R}^n$ be a linear operator all of whose eigenvalues have positive real parts. Then the system*

$$\dot{\mathbf{x}} = A\mathbf{x}, \qquad \mathbf{x} \in \mathbf{R}^n$$

is topologically equivalent to the standard system (Fig. 147)

$$\dot{\mathbf{x}} = \mathbf{x}, \qquad \mathbf{x} \in \mathbf{R}^n.$$

Proof. The lemma is almost obvious in the one-dimensional case and in the case of a focus in the plane, and hence, by Lemma 1, in any system without multiple roots. The proof of Lemma 3 in the general case will be given later. $\blacksquare$

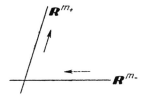

Fig. 146   Invariant subspaces of an operator with no purely imaginary eigenvalues.

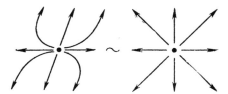

Fig. 147   All unstable nodes are topologically equivalent.

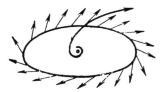

Fig. 148   Level surface of the Lyapunov function.

22.3. The Lyapunov function. The proof of Lemma 3 is based on the construction of a special quadratic form, called the *Lyapunov function*.

THEOREM. *Let* $A: \mathbf{R}^n \to \mathbf{R}^n$ *be a linear operator all of whose eigenvalues have positive real parts. Then there exists a Euclidean structure in* $\mathbf{R}^n$ *such that the vector* $A\mathbf{x}$ *makes an acute angle with the radius vector* $\mathbf{x}$ *at every point* $\mathbf{x} \neq 0$.

In other words:

There exists a positive definite quadratic form r^2 *in* $\mathbf{R}^n$ *such that its derivative in the direction of the vector field* $A\mathbf{x}$ *is positive*:

$$L_{A\mathbf{x}}r^2 > 0 \quad \forall\, \mathbf{x} \neq 0. \tag{1}$$

Or alternatively:

There exists an ellipsoid in $\mathbf{R}^n$ *with center* 0 *such that the vector* $A\mathbf{x}$ *is directed outward at every point* $\mathbf{x}$ *of the ellipsoid* (Fig. 148).

The equivalence of all three formulations is easily verified. We will prove (and subsequently use) the theorem in the second formulation. The proof is most convenient in the complex case:

Suppose all the eigenvalues λ_k *of the operator* $A: \mathbf{C}^n \to \mathbf{C}^n$ *have positive real parts. Then there exists a positive definite quadratic form* $r^2: {}^{\mathbf{R}}\mathbf{C}^n \to \mathbf{R}$ *whose derivative along the direction of the vector field* ${}^{\mathbf{R}}A\mathbf{z}$ *is a positive definite quadratic form*:

$$L_{{}^{\mathbf{R}}A\mathbf{z}}r^2 > 0 \quad \forall\, \mathbf{z} \neq 0. \tag{2}$$

Applying the inequality (2) in the case where the operator A is the complexification of a real operator and $\mathbf{z}$ belongs to a real subspace (Fig. 149), we get the real theorem (1).

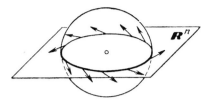

Fig. 149   Level surface of the Lyapunov function in $\mathbf{C}^n$.

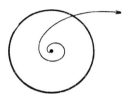

Fig. 150   Positive definiteness of the form (4) in the case $n = 1$.

22.4. Construction of the Lyapunov function. We will choose the Lyapunov function r^2 to be the sum of the squares of the moduli of the co-ordinates in a suitable complex basis:

$$r^2 = (\mathbf{z}, \bar{\mathbf{z}}) = \sum_{k=1}^{n} z_k \bar{z}_k.$$

In a fixed basis we can identify the vector $\mathbf{z}$ with a set of numbers $z_1, \ldots, z_n$ and the operator $A: \mathbf{C}^n \to \mathbf{C}^n$ with a matrix (a_{kl}). A calculation now shows that *the derivative is a quadratic form*:

$$L_{\mathbf{R}_{A\mathbf{z}}}(\mathbf{z}, \bar{\mathbf{z}}) = (A\mathbf{z}, \bar{\mathbf{z}}) + (\mathbf{z}, \overline{A\mathbf{z}}) = 2 \operatorname{Re} (A\mathbf{z}, \bar{\mathbf{z}}). \tag{3}$$

If the basis is an eigenbasis, this function is positive definite (Fig. 150). In fact, we then have

$$2 \operatorname{Re} (A\mathbf{z}, \bar{\mathbf{z}}) = 2 \sum_{k=1}^{n} \operatorname{Re} \lambda_k |z_k|^2. \tag{4}$$

But all the real parts of the eigenvalues λ_k are positive, by hypothesis, and hence the form (4) is positive definite.

If the operator A has no eigenbasis, then it has an "almost proper" basis which can be used with equal success to construct the Lyapunov function. More exactly, we have

LEMMA 4. *Let* $A: \mathbf{C}^n \to \mathbf{C}^n$ *be a* $\mathbf{C}$-*linear operator and let* $\varepsilon > 0$. *Then a basis* $\xi_1, \ldots, \xi_n$ *can be chosen in* $\mathbf{C}^n$ *in which the matrix of* A *has "upper-triangular" form*

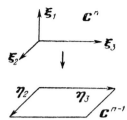

Fig. 151   Construction of a basis in which the matrix of the operator is triangular.

with all elements above the main diagonal of modulus less then ε:

$$(A) = \begin{pmatrix} \lambda_1 & & < \varepsilon \\ & \ddots & \\ 0 & & \lambda_n \end{pmatrix}.$$

Proof. The existence of a basis in which the matrix is upper-triangular follows, for example, from the theorem on the Jordan canonical form.

We can easily construct such a basis by induction in n, using only the fact that every linear operator $A: \mathbf{C}^n \to \mathbf{C}^n$ has an eigenvector. Let ξ_1 be this vector (Fig. 151), and consider the factor space $\mathbf{C}^n/\mathbf{C}\xi_1 \cong \mathbf{C}^{n-1}$. Then the operator A determines an operator $\tilde{A}: \mathbf{C}^{n-1} \to \mathbf{C}^{n-1}$ on the factor space. Let $\eta_2, \ldots, \eta_n$ be a basis in $\mathbf{C}^{n-1}$ in which the matrix of the operator $\tilde{A}$ is upper-triangular, and let $\xi_2, \ldots, \xi_n$ be any representatives of the classes $\eta_2, \ldots, \eta_n$ in $\mathbf{C}^n$. Then $\xi_1, \xi_2, \ldots, \xi_n$ is the desired basis.

Now suppose the matrix of the operator A is upper-triangular in the basis $\xi_1, \ldots, \xi_n$. Then *the elements above the main diagonal can be made arbitrarily small by replacing the basis vectors by proportional vectors.* In fact, let a_{kl} be the elements of the operator A in the basis ξ_k, so that $a_{kl} = 0$ if $k > l$. Then the elements of the matrix of A in the basis $\xi'_k = N^k \xi_k$ are just

$$a'_{kl} = \frac{a_{kl}}{N^{l-k}}.$$

But $|a'_{kl}| < \varepsilon$ for all $l > k$ if N is sufficiently large.   ∎

The sum of the squares of the moduli in this "ε-almost proper" basis will be chosen as the Lyapunov function (for sufficiently small ε).

Consider the set of all quadratic forms in $\mathbf{R}^n$. This set has the natural structure of a linear space $\mathbf{R}^{m(m+1)/2}$. Our next result is almost obvious:

LEMMA 5. *The set of positive definite quadratic forms in $\mathbf{R}^m$ is open in $\mathbf{R}^{m(m+1)/2}$. In other words, if a form*

$$a(\mathbf{x}) = \sum_{k,l=1}^{m} a_{kl} x_k x_l$$

is positive definite, then there exists an $\varepsilon > 0$ *such that every form* $a(\mathbf{x}) + b(\mathbf{x})$ *where* $|b_{kl}| < \varepsilon$ *(for all* $k, l, 1 \leqslant k, l \leqslant m$*) is also positive definite.*

Proof. The form $a(\mathbf{x})$ is positive at all points of the unit sphere

$$\sum_{k=1}^{m} x_k^2 = 1.$$

The sphere is compact, and the form is continuous. Therefore the greatest lower bound is achieved, so that $a(\mathbf{x}) \geqslant \alpha > 0$ everywhere on the sphere. If $|b_{kl}| < \varepsilon$, then

$$|b(\mathbf{x})| \leqslant \sum_{k,l=1}^{m} |b_{kl}| < m^2 \varepsilon$$

on the sphere. Therefore the form $a(\mathbf{x}) + b(\mathbf{x})$ is positive on the sphere if $\varepsilon < \alpha/m^2$, and hence is positive definite. ∎

Remark. Our argument also implies that every positive definite quadratic form $a(\mathbf{x})$ satisfies the inequality

$$\alpha|\mathbf{x}|^2 \leqslant a(\mathbf{x}) \leqslant \beta|\mathbf{x}|^2, \qquad 0 < \alpha < \beta \tag{5}$$

everywhere.

Problem 1. Prove that the set of nondegenerate quadratic forms with a given signature is open.

Example 1. The space of quadratic forms $ax^2 + 2bxy + cy^2$ in two variables is a three-dimensional space with coordinates a, b, c (Fig. 152). The case $b^2 = ac$ separates this space into three open parts according to the signature.

22.5. Proof of Theorem 22.3. Consider the derivative along the direction of the vector field $^{\mathbf{R}}A\mathbf{z}$ of the sum of the squares of the moduli of the co-ordinates in the "ε-almost proper" basis chosen in Lemma 4. According to (3), this derivative is a quadratic form in the real and imaginary parts of the coordinates $z_k = x_k + iy_k$. Separating the terms in (3) corresponding to the

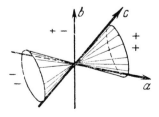

Fig. 152   The space of quadratic forms.

elements of the matrix (A) on the main diagonal from those corresponding to the elements of (A) above the main diagonal, we get

$$L_{\mathbf{R}_{Az}}r^2 = P + Q,$$

where

$$P = 2 \operatorname{Re} \sum_{k=l} a_{kl}z_k\bar{z}_l, \qquad Q = 2 \operatorname{Re} \sum_{k<l} a_{kl}z_k\bar{z}_l.$$

Since the diagonal elements of the triangular matrix (A) are just the eigenvalues λ_k of the operator A, *the quadratic form*

$$P = 2 \operatorname{Re} \sum_{k=1}^{n} \lambda_k(x_k^2 + y_k^2)$$

in the variables x_k, y_k is positive definite and independent of the choice of basis.† It follows from Lemma 5 that for sufficiently small ε, the form $P + Q$ (which is close to P) is also positive definite. In fact, for sufficiently small ε, the coefficients of the variables x_k, y_k in the form Q become arbitrarily small (since $|a_{kl}| < \varepsilon$ for $k < l$). But this implies (2) and hence (1). $\blacksquare$

Remark. Since $L_{A\mathbf{x}}r^2$ is a positive definite quadratic form, it satisfies an inequality of the type (5):

$$\alpha r^2 \leqslant L_{A\mathbf{x}}r^2 \leqslant \beta r^2, \qquad 0 < \alpha < \beta. \tag{5'}$$

The following series of problems leads to another proof of Theorem 2.3.

Problem 2. Prove that differentiation in the direction of the vector field $A\mathbf{x}$ in $\mathbf{R}^n$ gives a linear operator $L_A : \mathbf{R}^{n(n+1)/2} \to \mathbf{R}^{n(n+1)/2}$ from the space of quadratic forms on $\mathbf{R}^n$ into itself.

Problem 3. From a knowledge of the eigenvalues λ_i of the operator A, find the eigenvalues of the operator L_A.

Ans. $\lambda_i + \lambda_j$, $1 \leqslant i, j \leqslant n$.

Hint. Suppose A has an eigenbasis. Then the eigenvectors of L_A consist of the quadratic forms equal to products of pairs of linear forms which are eigenvectors of the operator dual to A.

Problem 4. Prove that the operator L_A is an isomorphism if A has no pair of eigenvalues λ, μ such that $\lambda = -\mu$. In particular, prove that if the real parts of all the eigenvalues of the operator A are of the same sign, then every quadratic form on $\mathbf{R}^n$ is the derivative of a quadratic form in the direction of the vector field $A\mathbf{x}$.

Problem 5. Prove that if the real parts of all the eigenvalues of an operator A are positive and if the derivative of a quadratic form in the direction of the field $A\mathbf{x}$ is positive definite, then the form is itself positive definite (and hence satisfies all the requirements of Theorem 22.3).

† It should be noted that the *mapping* $\mathbf{RC}^n \to \mathbf{R}$ specified by the form P *does depend* on the choice of basis.

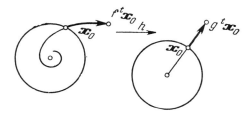

Fig. 153   Construction of the homeomorphism h.

Hint. Represent the form as the integral of its derivative along the phase curves.

22.6. Construction of the homeomorphism h. In order to prove Lemma 3, we now construct a homeomorphism $h\colon \mathbf{R}^n \to \mathbf{R}^n$ carrying the phase flow $\{f^t\}$ of the equation $\dot{\mathbf{x}} = A\mathbf{x}$ (Re $\lambda_k > 0$) into the phase flow $\{g^t\}$ of the equation $\dot{\mathbf{x}} = \mathbf{x}$. Let S be the sphere (or ellipsoid)

$$S = \{\mathbf{x} \in \mathbf{R}^n \colon r^2(\mathbf{x}) = 1\},$$

where r^2 is the Lyapunov function of Sec. 22.3, and let h be such that
i) The points of S are invariant under h;
ii) If $\mathbf{x}_0$ is a point of S, then h carries the point $f^t\mathbf{x}_0$ of the phase trajectory of the equation $\dot{\mathbf{x}} = A\mathbf{x}$ into the point $g^t\mathbf{x}_0$ of the phase trajectory of the equation $\dot{\mathbf{x}} = \mathbf{x}$ (Fig. 153):

$$\begin{cases} h(f^t\mathbf{x}_0) = g^t\mathbf{x}_0 & \forall\, t \in \mathbf{R},\ \mathbf{x}_0 \in S, \\ h(0) = 0. \end{cases} \tag{6}$$

We must now verify the following facts, whose proofs are almost obvious:
1) Formula (6) uniquely defines the value of h at every point $\mathbf{x} \in \mathbf{R}^n$;
2) The mapping $h\colon \mathbf{R}^n \to \mathbf{R}^n$ is one-to-one and continuous in both directions;
3) $h \circ f^t = g^t \circ h$.

22.7. Proof of Lemma 3. First we prove

LEMMA 6. *Let* $\varphi\colon R^n \to R^n$ *be any solution of the equation* $A\dot{\mathbf{x}} = \mathbf{x}$ *distinct from zero, and form the real function*

$$\rho(t) = \ln r^2(\varphi(t))$$

of the real variable t. *Then the mapping* $\rho\colon R \to R$ *is a diffeomorphism, and*

$$\alpha \leqslant \frac{d\rho}{dt} \leqslant \beta. \tag{7}$$

Proof. By the uniqueness theorem, we have

$$r^2(\varphi(t)) \neq 0 \quad \forall\, t \in \mathbf{R}.$$

Moreover

$$\frac{d\rho}{dt} = \frac{L_{A\mathbf{x}}r^2}{r^2}$$

satisfies the estimate (7) because of (5′). ∎

COROLLARY 1. *Every point* $\mathbf{x} \neq 0$ *can be represented in the form*

$$\mathbf{x} = f^t\mathbf{x}_0, \tag{8}$$

where $\mathbf{x}_0 \in S$, $t \in \mathbf{R}$ *and* $\{f^t\}$ *is the phase flow of the equation* $\dot{\mathbf{x}} = A\mathbf{x}$.

Proof. Consider the solution $\boldsymbol{\varphi}$ with initial condition $\boldsymbol{\varphi}(0) = \mathbf{x}$. By Lemma 6, $r^2(\boldsymbol{\varphi}(\tau)) = 1$ for some τ. The point $\mathbf{x}_0 = \boldsymbol{\varphi}(\tau)$ belongs to the sphere S. Setting $t = -\tau$, we get $\mathbf{x} = f^t\mathbf{x}_0$. ∎

COROLLARY 2. *The representation* (8) *is unique.*

Proof. The phase curve leaving $\mathbf{x}$ (Fig. 153) is unique and intersects the sphere in a single point $\mathbf{x}_0$, by Lemma 6. The uniqueness of t follows from the monotonicity of $\rho(t)$, again by Lemma 6. ∎

Thus we have constructed a one-to-one mapping of the direct product of the line and the sphere onto Euclidean space minus a single point:

$$F: \mathbf{R} \times S^{n-1} \to \mathbf{R}^n\backslash 0, \qquad F(t, \mathbf{x}_0) = f^t\mathbf{x}_0.$$

It follows from the theorem on the dependence of the solution on the initial conditions that F, as well as the inverse mapping F^{-1}, is continuous (and even a diffeomorphism).

We now note that

$$\frac{d\rho}{dt} = 2$$

for the standard equation $\dot{\mathbf{x}} = \mathbf{x}$. Hence the mapping

$$G: \mathbf{R} \times S^{n-1} \to \mathbf{R}^n\backslash 0, \qquad G(t, \mathbf{x}_0) = g^t\mathbf{x}_0$$

is also one-to-one and continuous in both directions. According to the definition (6), the mapping h coincides with the mapping $G \circ F^{-1}: \mathbf{R}^n\backslash 0 \to \mathbf{R}^n\backslash 0$ everywhere except at the point 0. Thus we have proved that $h: \mathbf{R}^n \to \mathbf{R}^n$ is a one-to-one mapping.

The continuity of h and h^{-1} everywhere except at the point 0 follows from the continuity of F, F^{-1} and G, G^{-1}; actually h is a diffeomorphism everywhere except at the point 0 (Fig. 154). The continuity of h and h^{-1} at 0 follows from Lemma 6. This lemma even allows us to obtain an explicit estimate for $r^2(h(\mathbf{x}))$ in terms of $r^2(\mathbf{x})$, $|\mathbf{x}| \leqslant 1$:

$$(r^2(\mathbf{x}))^{2/\alpha} \leqslant r^2(h(\mathbf{x})) \leqslant (r^2(\mathbf{x}))^{2/\beta}.$$

In fact, let $\mathbf{x} = F(t, \mathbf{x}_0)$, $t \leqslant 0$. Then $\beta t \leqslant \ln r^2(\mathbf{x}) \leqslant \alpha t$ and $\ln r^2(h(\mathbf{x})) = 2t$. Moreover, for $\mathbf{x} \neq 0$ we have $\mathbf{x} = f^s\mathbf{x}_0$, and hence

$$\begin{aligned}(h \circ f^t)(\mathbf{x}) &= h(f^t(f^s\mathbf{x}_0)) = h(f^{t+s}\mathbf{x}_0) = g^{t+s}\mathbf{x}_0 \\ &= g^t(g^s\mathbf{x}_0) = g^t(h(\mathbf{x})) = (g^t \circ h)(\mathbf{x}),\end{aligned}$$

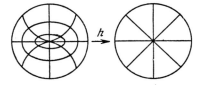

Fig. 154   The homeomorphism h is a diffeomorphism everywhere except at 0.

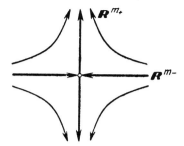

Fig. 155   The standard saddle.

while for $\mathbf{x} = 0$ we also have $(h \circ f^t)(\mathbf{x}) = (g^t \circ h)(\mathbf{x})$. Thus the validity of assertions 1)–3) of Sec. 22.6 has finally been established, and the proof of Lemma 3 is now complete. ∎

22.8. Proof of Theorem 22.1. It follows from Lemmas 1, 2, and 3 that every linear system $\dot{\mathbf{x}} = A\mathbf{x}$, where the operator $A : \mathbf{R}^n \to \mathbf{R}^n$ has no purely imaginary eigenvalues, is topologically equivalent to the standard multi-dimensional "saddle"

$$\begin{cases} \dot{\mathbf{x}}_1 = -\mathbf{x}_1, & \mathbf{x}_1 \in \mathbf{R}^{m_-}, \\ \dot{\mathbf{x}}_2 = \mathbf{x}_2, & \mathbf{x}_2 \in \mathbf{R}^{m_+} \end{cases}$$

(Fig. 155). Hence two such systems with identical numbers m_- and m_+ are topologically equivalent (to each other). Note that the subspaces $\mathbf{R}^{m_-}$ and $\mathbf{R}^{m_+}$ are invariant under the phase flow $\{g^t\}$. As t increases, every point of $\mathbf{R}^{m_-}$ approaches 0.

Problem 1. Prove that $g^t\mathbf{x} \to 0$ as $t \to +\infty$ if and only if $\mathbf{x} \in \mathbf{R}^{m_-}$.

Therefore $\mathbf{R}^{m_-}$ is called the *incoming strand* of the saddle. In just the same way, $\mathbf{R}^{m_+}$ is called the *outgoing strand*, defined by the condition $g^t\mathbf{x} \to 0$ as $t \to -\infty$.

We now prove the second part of Theorem 22.1, namely, *two topologically equivalent systems have the same number of eigenvalues with negative real parts.* This number is just the dimension m_- of the incoming strand, and hence we need only show that *the dimensions of the incoming strands of two topologically equivalent saddles are identical.*

First we note that every homeomorphism h carrying the phase flow of one saddle into the phase flow of another must carry the incoming strand of one saddle into the incoming strand of the other (since approach to 0 as $t \to +\infty$ is preserved under a homeomorphism). Hence the homeomorphism h also establishes a homeomorphic mapping of the incoming strand of one saddle onto the incoming strand of the other.

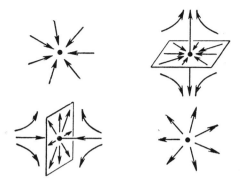

Fig. 156   Strands of three-dimensional saddles.

The fact that the strands have the same dimension now follows from the following key proposition of topology:

The dimension of the space $\mathbf{R}^n$ is a topological invariant, i.e., a homeomorphism $h: \mathbf{R}^m \to \mathbf{R}^n$ can exist only between spaces of the same dimension.†

Although this proposition seems "obvious," its proof is not easy and will not be given here.

Problem 2. Prove that the 4 saddles with a three-dimensional phase space such that $(m_-, m_+) = (3, 0), (2, 1), (1, 2), (0, 3)$ are topologically nonequivalent (without using the unproved topological proposition).

Hint. A one-dimensional strand consists of three phase curves, while a multidimensional strand consists of infinitely many phase curves (Fig. 156).

Thus for $\mathbf{R}^1$, $\mathbf{R}^2$, $\mathbf{R}^3$ we have completely proved the topological classification of linear systems whose eigenvalues have nonzero real parts. However for $\mathbf{R}^n$, $n > 3$ we are compelled to refer to the above unproved proposition on the topological invariance of dimension. ∎

Problem 3. Carry out the topological classification of linear operators $A: \mathbf{R}^n \to \mathbf{R}^n$ with no eigenvalues of modulus 1. Show that the unique topological invariant is the number of eigenvalues of modulus less than 1.

23. Stability of Equilibrium Positions

The problem of stability of an equilibrium position of a nonlinear system is solved in the same way as for a linearized system, provided the latter has no eigenvalues on the imaginary axis.

† However there exist one-to-one mappings $\mathbf{R}^m \to \mathbf{R}^n$, as well as continuous mappings of $\mathbf{R}^m$ *onto* $\mathbf{R}^n$ with $m < n$ (for example $\mathbf{R}^1 \to \mathbf{R}^2$).

23.1. Lyapunov stability. Consider the equation

$$\dot{\mathbf{x}} = \mathbf{v}(\mathbf{x}), \qquad \mathbf{x} \in U \subset \mathbf{R}^n, \tag{1}$$

where $\mathbf{v}$ is a vector field differentiable $r > 2$ times in the domain U. Suppose equation (1) has an equilibrium position (Fig. 157), and choose coordinates x_i such that the equilibrium position is at the origin: $\mathbf{v}(0) = 0$. The solution with initial condition $\boldsymbol{\varphi}(t_0) = 0$ is just $\boldsymbol{\varphi} = 0$, and we are interested in the behavior of solutions with neighboring initial conditions.

Definition. An equilibrium position $\mathbf{x} = 0$ of equation (1) is said to be *stable* (*in Lyapunov's sense*) if given any $\varepsilon > 0$, there exists a $\delta > 0$ (depending only on ε and not on t, about which more later) such that for every $\mathbf{x}_0$ for which $|\mathbf{x}_0| < \delta$,† the solution $\boldsymbol{\varphi}$ of (1) with initial condition $\boldsymbol{\varphi}(0) = \mathbf{x}_0$ can be extended onto the whole half-line $t > 0$ and satisfies the inequality $|\boldsymbol{\varphi}(t)| < \varepsilon$ for all $t > 0$ (Fig. 158).

Problem 1. Investigate the Lyapunov stability of the equilibrium positions of the following equations:

a) $\dot{x} = 0$;
b) $\dot{x} = x$;

c) $\begin{cases} \dot{x}_1 = x_2, \\ \dot{x}_2 = -x_1; \end{cases}$

d) $\begin{cases} \dot{x}_1 = x_1, \\ \dot{x}_2 = -x_2; \end{cases}$

e) $\begin{cases} \dot{x}_1 = x_2, \\ \dot{x}_2 = -\sin x_1. \end{cases}$

Fig. 157   Do the phase curves starting in a sufficiently small neighborhood of an equilibrium position stay near the equilibrium position?

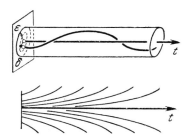

Fig. 158   Difference in behavior of integral curves for stable and unstable equilibrium positions.

† As usual, $|\mathbf{x}|^2 = x_1^2 + \cdots + x_n^2$ if $\mathbf{x} = (x_1, \ldots, x_n)$.

Fig. 159   Integral curves for an asymptotically stable equilibrium position.

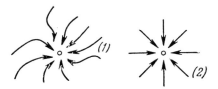

Fig. 160   Phase curves of equations (1) and (2).

Problem 2. Prove that the above definition is correct, i.e., that the stability of the equilibrium position is independent of the system of coordinates figuring in the definition.

Problem 3. Suppose it is known that for any $N > 0$, $\varepsilon > 0$, there exists a solution φ of equation (1) such that $|\varphi(0)| < \varepsilon$ and $|\varphi(t)| > N|\varphi(0)|$ for some $t > 0$. Does this imply that the equilibrium position $\mathbf{x} = 0$ is unstable?

23.2. Asymptotic stability.

Definition. An equilibrium position $\mathbf{x} = 0$ of equation (1) is said to be *asymptotically stable* if it is stable in Lyapunov's sense and if

$$\lim_{t \to +\infty} \varphi(t) = 0$$

for every solution $\varphi(t)$ with an initial condition $\varphi(0)$ lying in a sufficiently small neighborhood of zero (Fig. 159).

Problem 1. Investigate the asymptotic stability of the equilibrium positions of the following equations:

a) $\dot{x} = 0;$ c) $\begin{cases} \dot{x}_1 = x_2, \\ \dot{x}_2 = -x_1. \end{cases}$
b) $\dot{x} = x;$

Problem 2. Suppose every solution approaches the equilibrium position as $t \to +\infty$. Does this imply Lyapunov stability of the equilibrium position?

23.3. Stability in terms of the behavior of the first approximation.
Together with equation (1), we now consider the linearized equation (Fig. 160)

$$\dot{\mathbf{x}} = A\mathbf{x}, \qquad A: \mathbf{R}^n \to \mathbf{R}^n. \tag{2}$$

Then $\mathbf{v} = \mathbf{v}_1 + \mathbf{v}_2$, where

$$\mathbf{v}_1(x) = A\mathbf{x}, \qquad \mathbf{v}_2 = O(|\mathbf{x}|^2).$$

Fig. 161   Eigenvalues of the operator A.

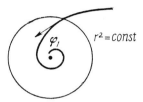

Fig. 162   Level surface of the Lyapunov function.

THEOREM. *Suppose all the eigenvalues λ_k of the operator A lie in the left half-plane* Re $\lambda < 0$ (Fig. 161). *Then the equilibrium position $\mathbf{x} = 0$ of equation* (1) *is asymptotically stable.*

Problem 1. Give an example of an unstable equilibrium position (in the sense of Lyapunov) of equation (1) for which all Re $\lambda_k \leqslant 0$.

Comment. It can be shown that if the real part of at least one eigenvalue is *positive*, then the equilibrium position is unstable. In the case of zero real parts, the stability depends on terms of the Taylor series of order higher than 1.

Problem 2. Is the zero equilibrium position of the system

$$\begin{cases} \dot{x}_1 = x_2, \\ \dot{x}_2 = -x_1^n \end{cases}$$

stable (both in Lyapunov's sense and asymptotically)?

Ans. If n is even, it is unstable (in Lyapunov's sense), while if n is odd, it is stable in Lyapunov's sense but not asymptotically.

23.4. Proof of Theorem 23.3. According to Sec. 22.3, there exists a Lyapunov function, i.e., a positive definite quadratic form r^2 whose derivative in the direction of the linear field $\mathbf{v}_1$ is negative definite, so that

$$L_{\mathbf{v}_1} r^2 \leqslant -2\gamma r^2$$

where γ is a positive constant (Fig. 162).

LEMMA. *The derivative of the Lyapunov function in the direction of the nonlinear field $\mathbf{v}$ satisfies the inequality*

$$L_{\mathbf{v}} r^2 \leqslant -\gamma r^2 \tag{3}$$

in a sufficiently small neighborhood of the point $\mathbf{x} = 0$.

Proof. Clearly

$$L_{\mathbf{v}}r^2 = L_{\mathbf{v}_1}r^2 + L_{\mathbf{v}_2}r^2.$$

But

$$L_{\mathbf{v}_2}r^2 = O(r^3), \tag{4}$$

so that the second term is much smaller than the first if r is small. In fact,

$$L_{\mathbf{u}}f = \sum_{i=1}^{n} \frac{\partial f}{\partial x_i}u_i$$

for any field $\mathbf{u}$ and any function f, where in our case $\mathbf{u} = \mathbf{v}_2, f = r^2, u_i = O(r^2)$ and $\partial f/\partial x_i = O(r)$ (why?), which implies (4).

Thus there exist constants $C > 0$, $\sigma > 0$ such that

$$|L_{\mathbf{v}}r^2|_{\mathbf{x}} \leqslant C|r^2(\mathbf{x})|^{3/2}$$

for all $\mathbf{x}$ with $|\mathbf{x}| < \sigma$. The right-hand side is no larger than γr^2 for sufficiently small $|\mathbf{x}|$, and hence

$$L_{\mathbf{v}}r^2 \leqslant -2\gamma r^2 + \gamma r^2 = -\gamma r^2$$

in a neighborhood of the point $\mathbf{x} = 0$. ∎

Proof of Theorem 23.3.   Let φ be a nonzero solution of equation (1) satisfying the initial condition $\mathbf{x} = 0$ in a sufficiently small neighborhood of the point $\mathbf{x} = 0$, and consider the following function of time:

$$\rho(t) = \ln r^2(\varphi(t)), \qquad t \geqslant 0.$$

By the uniqueness theorem $r^2(\varphi(t)) \neq 0$, so that the function φ is defined and differentiable. According to the inequality (3), we have

$$\dot\rho = \frac{1}{r^2 \circ \varphi}\frac{d}{dt}r^2 \circ \varphi = \frac{L_{\mathbf{v}}r^2}{r^2} \leqslant -\gamma.$$

It follows that $r^2(\varphi(t))$ decreases monotonically and approaches 0 as $t \to +\infty$:

$$\rho(t) \leqslant \rho(0) - \gamma t,$$
$$r^2(\varphi(t)) \leqslant r^2(\varphi(0))e^{-\gamma t} \to 0. \quad ∎ \tag{5}$$

Problem 1.   Find the gap in this proof.

Ans.   We did not prove that the solution φ can be extended indefinitely forward.

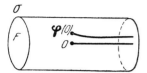

Fig. 163   The solution can be extended forward indefinitely.

Completion of the proof.   Let $\sigma > 0$ be such that the inequality (3) holds for $|\mathbf{x}| < \sigma$, and consider the compact set

$$F = \{\mathbf{x}, t : r^2(\mathbf{x}) \leqslant \sigma, |t| \leqslant T\}$$

in extended phase space (Fig. 163). Let $\boldsymbol{\varphi}$ be the solution with initial condition $\boldsymbol{\varphi}(0)$, where $r^2(\boldsymbol{\varphi}(0)) < \sigma$. By the extension theorem, we can extend $\boldsymbol{\varphi}$ forward up to the boundary of the cylinder F. But the derivative of the function $r^2\boldsymbol{\varphi}(t)$ is negative as long as the point $(t, \boldsymbol{\varphi}(t))$ belongs to F. Therefore the solution cannot leave the lateral surface of the cylinder F (where $r^2 = \sigma^2$), and hence it can be extended up to the end face of the cylinder (where $t = T$). Therefore, since T is arbitrary (and independent of σ), the solution $\boldsymbol{\varphi}$ can be extended indefinitely forward. Moreover. $r^2(\boldsymbol{\varphi}(t)) < \sigma^2$ and the inequality (3) holds for all $t \geqslant 0$. ∎

Remark 1.   Actually we have proved more than the asymptotic stability of the equilibrium position. In fact, it is clear from the inequality (5) that the convergence $\boldsymbol{\varphi}(t) \to 0$ is uniform (with respect to initial conditions $\mathbf{x}_0$ sufficiently close to 0). Moreover, (5) shows the rate of convergence (namely exponential).

In essence, Theorem 23.3 asserts that the uniform convergence to 0 of the solutions of the linear equation (2) is not destroyed by a nonlinear perturbation $\mathbf{v}_2(\mathbf{x}) = O(|\mathbf{x}|^2)$ of the right-hand side of the equation. A similar assertion is valid for various perturbations of a more general nature. For example, we might consider a nonautonomous perturbation $\mathbf{v}_2(\mathbf{x}, t)$ such that $|\mathbf{v}_2(\mathbf{x}, t)| \leqslant \psi(|\mathbf{x}|)$ where $\psi(|\mathbf{x}|) = o(|\mathbf{x}|)$ as $\mathbf{x} \to 0$.

Problem 2. Prove that under the conditions of the theorem, equations (1) and (2) are topologically equivalent in a neighborhood of the equilibrium position.

Remark 2.   Theorem 23.3 leads to the following algebraic problem, known as the *Routh-Hurwitz problem: Determine whether or not all the zeros of a given polynomial lie in the left half-plane.* This problem can be solved by a finite number of arithmetic operations on the coefficients of the polynomial. The appropriate algorithms are described in courses on algebra (Hurwitz's criterion, Sturm's method) and complex analysis (the argument principle, the methods of Vyshnegradski, Nyquist, and Mikhailov).† We will return to the Routh-Hurwitz problem in Sec. 36.4.

24. The Case of Purely Imaginary Eigenvalues

Linear equations with no purely imaginary eigenvalues have been investigated in detail (Secs. 21 and 22), and their phase curves behave rather simply, exhibiting a saddle (Sec. 22.8). We now turn to the case of linear equations with purely imaginary eigenvalues, whose phase curves offer examples of more complicated behavior. Such equations are encountered, for example, in the theory of oscillations of conservative systems (see Sec. 25.6).

24.1. Topological classification. Suppose all the eigenvalues $\lambda_1, \ldots, \lambda_n$ of the linear equation

$$\dot{\mathbf{x}} = A\mathbf{x}, \qquad \mathbf{x} \in \mathbf{R}^n, \quad A : \mathbf{R}^n \to \mathbf{R}^n \tag{1}$$

are purely imaginary. Then under what conditions are two such equations topologically equivalent? The answer to this question is not known, and evidently the problem cannot be solved by presently available mathematical methods.

Problem 1. Prove that in the case of the plane ($n = 2$, $\lambda_{1,2} = \pm i\omega \neq 0$), algebraic equivalence (i.e., equality of eigenvalues) is a necessary and sufficient condition for topological equivalence.

24.2. Example. Consider the equation

$$\begin{cases} \dot{x}_1 = \omega_1 x_1, \\ \dot{x}_2 = -\omega_1 x_1, \\ \dot{x}_3 = \omega_2 x_4, \\ \dot{x}_4 = -\omega_2 x_3 \end{cases} \qquad \begin{aligned} \lambda_{1,2} &= \pm i\omega_1, \\[2mm] \lambda_{3,4} &= \pm i\omega_2 \end{aligned} \tag{2}$$

in $\mathbf{R}^4$. The space $\mathbf{R}^4$ decomposes into a direct sum $\mathbf{R}^4 = \mathbf{R}_{1,2} + \mathbf{R}_{3,4}$ of two planes (Fig. 164), and correspondingly the system (2) decomposes into two independent systems

---

† See e.g., A. G. Kurosh, *A. Course in Higher Algebra* (in Russian), Moscow (1968), Chap. 9; M. A. Lavrentev and B. V. Shabat, *Methods of the Theory of Functions of a Complex Variable* (in Russian), Moscow (1958), Chap. 5; N. G. Chebotarev and N. N. Meiman, *The Routh-Hurwitz Problem for Polynomials and Entire Functions* (in Russian), Trudy Mat. Inst. Steklov, Moscow (1949), No. XXVI.

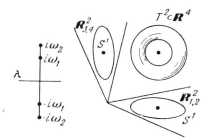

Fig. 164   Phase space of the system (2).

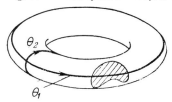

Fig. 165   A torus.

$$\begin{cases} \dot{x}_1 = \omega_1 x_2, \\ \dot{x}_2 = -\omega_1 x_1, \end{cases} \quad (x_1, x_2) \in \mathbf{R}_{1,2},$$

$$\begin{cases} \dot{x}_3 = \omega_2 x_4, \\ \dot{x}_4 = -\omega_2 x_3, \end{cases} \quad (x_3, x_4) \in \mathbf{R}_{3,4}. \tag{3}$$

In each of these planes, the phase curves are circles, say

$$S^1 = \{x \in \mathbf{R}_{1,2}: x_1^2 + x_2^2 = C > 0\}$$

or points $(C = 0)$, and the phase flow consists of rotations (through angles $\omega_1 t$ and $\omega_2 t$ respectively).

   Every phase curve of equation (2) belongs to the direct product of the phase curves in the planes $\mathbf{R}_{1,2}$ and $\mathbf{R}_{3,4}$. Suppose the two curves are circles, with the direct product

$$T^2 = S^1 \times S^1 = \{\mathbf{x} \in \mathbf{R}^4: x_1^2 + x_2^2 = C, x_3^2 + x_4^2 = D\},$$

called a *two-dimensional torus*. To better visualize the torus T^2, we can proceed as follows. Consider the surface of the doughnut in $\mathbf{R}^3$ (Fig. 165) obtained by rotating a circle about an axis which lies in its plane but does not intersect it. A point of this surface is specified by two angular coordinates $\theta_1, \theta_2 \bmod 2\pi$, called the *longitude* and the *latitude* for a reason which is apparent from the figure. The coordinates θ_1 and θ_2 give a diffeomorphism of the surface of the doughnut and the direct product T^2 of the two circles.

The square $0 \leqslant \theta_1 \leqslant 2\pi,\ 0 \leqslant \theta_2 \leqslant 2\pi$ in the plane of the coordinates θ_1, θ_2 can be regarded as a map of the torus T^2 (Fig. 166) if we "paste together" every pair of points $(\theta_1, 0)$ and $(\theta_1, 2\pi)$ as well as every pair of points $(0, \theta_2)$ and $(2\pi, \theta_2)$. The whole plane can also be regarded as the map, but then every point of the torus has infinitely many images on the map.

The torus $T^2 \subset \mathbf{R}^4$ is invariant under the phase flow of equation (2), and the phase curves of (2) lie on the surface T^2. If θ_1 is the polar angle in the plane $\mathbf{R}_{1,2}$ measured from the direction of x_2 to that of x_1, then, according to (3), $\dot{\theta}_1 = \omega_1$. Similarly, measuring θ_2 from x_4 to x_3, we get $\dot{\theta}_2 = \omega_2$. Thus *the phase trajectories of the flow* (2) *on the surface* T^2 *satisfy the differential equation*

$$\dot{\theta}_1 = \omega_1, \qquad \dot{\theta}_2 = \omega_2, \tag{4}$$

so that the longitude and latitude of the phase point both change uniformly. This motion corresponds to that of a point "winding" around the torus (Fig. 167), represented by a straight line on the map of the torus.

24.3. Phase curves of equation (4) on the torus. Two numbers ω_1 and ω_2 are said to be *rationally independent* if

$$k_1\omega_1 + k_2\omega_2 = 0 \qquad (k_1, k_2 \text{ integral})$$

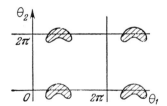

Fig. 166   A map of the torus.

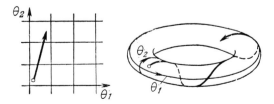

Fig. 167   A point winding around the torus.

implies $k_1 = k_2 = 0$. For example, $\sqrt{2}$ and $\sqrt{8}$ are rationally dependent, but not $\sqrt{6}$ and $\sqrt{8}$.

THEOREM.   *If ω_1 and ω_2 are rationally dependent, then every phase curve of equation (4) on the torus is closed. However, if ω_1 and ω_2 are rationally independent, then every phase curve of equation (4) is everywhere dense† on the torus T^2 (Fig. 168).*

In other words, suppose that every square of an infinite chessboard is occupied by a single (identically placed) rabbit, and suppose a hunter shoots in a direction whose angle of inclination with the lines of the chessboard has an irrational tangent. Then the hunter will hit at least one rabbit. (It is clear that if the tangent of the angle of inclination is rational, then we can place sufficiently small rabbits on the chessboard in such a way that the hunter will miss.)

LEMMA.   *Suppose the circle S^1 is rotated through an angle α which is incommensurable with 2π (Fig. 169). Then the images*

$$\theta, \quad \theta + \alpha, \quad \theta + 2\alpha, \quad \theta + 3\alpha, \dots \quad (\text{mod } 2\pi) \tag{5}$$

of any point on the circle under repeated application of the rotation form a set which is everywhere dense on the circle.

Proof.   The theorem can be deduced from the structure of the closed sub-

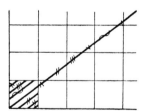

Fig. 168   An everywhere dense curve on the torus.

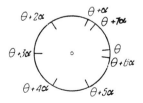

Fig. 169   Images of a point of the circle under repeated application of a rotation through the angle α.

† A set A is said to be *everywhere dense* in a space B if there is at least one point of A in an arbitrarily small neighborhood of every point of B.

groups of the line (see Sec. 10), but we will prove it from scratch, starting from the simple combinatorial fact that *if k + 1 objects are placed in k cells, then at least one cell contains more than one object* ("Dirichlet's cell principle"). Suppose we divide the circle into k equal half-open intervals of length $2\pi/k$. Then among the first $k + 1$ points of the sequence (5), there are two points in the same half-open interval. Let these points be $\theta + p\alpha$ and $\theta + q\alpha$ ($p > q$), and let $s = p - q$. Then the angle of rotation $s\alpha$ differs from a multiple of 2π by less than $2\pi/k$, and any two consecutive points of the sequence

$$\theta, \quad \theta + s\alpha, \quad \theta + 2s\alpha, \quad \theta + 3s\alpha, \ldots \quad (\text{mod } 2\pi) \tag{6}$$

(Fig. 170) are the same distance d apart, where $d < 2\pi/k$. Hence any ε-neighborhood of any point of S^1 contains points of the sequence (6), provided only that we choose k large enough to make $2\pi/k < \varepsilon$. ∎

Remark. We did not use the fact that α is incommensurable with 2π, but it is obvious that the lemma is false if α is commensurable with 2π.

Problem 1. Find and eliminate the gap in the proof of the theorem.

Proof of the theorem. The solution of equation (4) is of the form

$$\theta_1(t) = \theta_1(0) + \omega_1 t, \qquad \theta_2(t) = \theta_2(0) + \omega_2 t. \tag{7}$$

Suppose ω_1 and ω_2 are rationally dependent, so that

$$k_1\omega_1 + k_2\omega_2 = 0, \qquad k_1^2 + k_2^2 \neq 0.$$

Then the equations in T

$$\omega_1 T = 2\pi k_2, \qquad \omega_2 = -2\pi k_1$$

are compatible, and their solution gives the period of the closed phase curve (7). On the other hand, suppose ω_1 and ω_2 are rationally independent. Then ω_1/ω_2 is an irrational number. Consider the consecutive points of intersection of the phase curve (7) with the meridian $\theta_1 = 0 \pmod{2\pi}$. The latitudes of the points are

$$\theta_{2k} = \theta_{20} + 2\pi\frac{\omega_2}{\omega_1}k \qquad (\text{mod } 2\pi)$$

(Fig. 171). By the lemma, the set of points of intersection is everywhere dense on the meridian. But if L is a straight line in the plane and if we draw straight lines through a set of points everywhere dense in L in a direction different from the direction of L, then the lines form a set which is everywhere dense in the plane. It follows that the image†

---

† Here $[x]$ denotes the *integral part* of x, i.e., the largest integer $\leqslant x$.

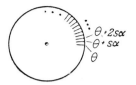

Fig. 170   The points $\theta + n s \alpha, n = 1, 2, \ldots$

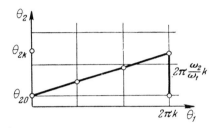

Fig. 171   Reduction of the theorem to the lemma.

$$\tilde{\theta}_1(t) = \theta_1(t) - 2\pi\left[\frac{\theta_1(t)}{2\pi}\right],$$

$$\tilde{\theta}_2(t) = \theta_2(t) - 2\pi\left[\frac{\theta_2(t)}{2\pi}\right]$$

of the phase curve (7) on the square $0 \leqslant \tilde{\theta}_1 < 2\pi, 0 \leqslant \tilde{\theta}_2 < 2\pi$ is everywhere dense. Therefore the phase curve of equation (4), and hence of equation (2), is everywhere dense on the torus.   ∎

The following problems give a number of simple implications of Theorem 2.3 outside the theory of ordinary differential equations.

Problem 1. Consider the sequence

1, 2, 4, 8, 1, 3, 6, 1, 2, 5, 1, 2, 4, 8, ...

of first digits of consecutive powers of 2. Does a 7 ever appear in this sequence? More generally, does 2^n begin with an arbitrary combination of digits?

Problem 2. Prove that

$$\sup_{0 < t < \infty} (\cos t + \sin \sqrt{2}t) = 2.$$

Problem 3. Find all closed subgroups of the group S^1 of complex numbers of modulus 1.

Ans. 1, S^1, $\{\sqrt[n]{1}\}$.

24.4. The multidimensional case.   Suppose the eigenvalues of equation (1) in $\mathbf{R}^{2m}$ are all simple, of the form

$$\lambda = \pm i\omega_1, \pm i\omega_2, \ldots, \pm i\omega_m.$$

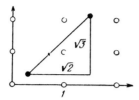

Fig. 172   A phase curve of the system $\dot\theta = 1$, $\dot\theta = \sqrt{2}$, $\dot\theta = \sqrt{3}$ is everywhere dense on the three-dimensional torus.

Then, arguing as in Sec. 24.2, we can show that the phase curves lie on the m-dimensional torus

$$T^m = S^1 \times \cdots \times S^1 = \{(\theta_1, \ldots, \theta_m) \bmod 2\pi\} \cong \mathbf{R}^m/Z^m$$

and satisfy the equations

$$\dot\theta_1 = \omega_1, \qquad \dot\theta_2 = \omega_2, \ldots, \qquad \dot\theta_m = \omega_m.$$

The numbers $\omega_1, \ldots, \omega_m$ are said to be *rationally independent* if

$$k_1\omega_1 + \cdots + k_m\omega_m = 0 \qquad (k_1, \ldots, k_m \text{ integral})$$

implies $k_1 = \cdots = k_m = 0$.

Problem 1. Prove that if the frequencies $\omega_1, \ldots, \omega_m$ are rationally independent, then every phase curve of equation (1) lying on the torus T^m is everywhere dense in T^m.

Corollary. Suppose a horse makes jumps $(\sqrt{2}, \sqrt{3})$ on a field where corn is planted in the pattern of a square grid (Fig. 172). Then the horse is certain to knock down at least one plant.

24.5. Uniform distribution.
The everywhere dense curves considered above have the remarkable property of being "uniformly distributed" on the surface of a torus. We now formulate the appropriate theorem in the simplest case. A sequence of points $\theta_1, \theta_2, \ldots$ on the circle $S^1 = \{\theta \bmod 2\pi\}$ is said to be *uniformly distributed* if given any arc $\Delta \subset S^1$, the number $N(\Delta, k)$ of points of the "initial section" $\theta_1, \ldots, \theta_k$ of the sequence which belong to Δ is asymptotically proportional to the length of Δ, i.e., if

$$\lim_{k \to \infty} \frac{N(\Delta, k)}{k} = \frac{|\Delta|}{2\pi}.$$

Problem 1. Prove that the sequence θ, $\theta + \alpha$, $\theta + 2\alpha$ is uniformly distributed on S^1 if the angle α is incommensurable with 2π.

Corollary. The numbers 2^n begin more often with 7 than with 8. Suppose $N_7(k)$ of the numbers $1, 2, 4, \ldots, 2^k$ begin with 7, while $N_8(k)$ begin with 8. Then the limit

$$\lim_{k \to \infty} \frac{N_7(k)}{N_8(k)}$$

exists.

Problem 2. Find this limit and show that it is greater than 1.

Comment. The initial section of the sequence (Sec. 24.3, Prob. 1) indicates that there are fewer sevens. This is due to the fact that the irrational number $\log_{10} 2 = 0.3010 \ldots$ is very close to the rational number $3/10$.

25. The Case of Multiple Eigenvalues

The solution of a linear equation with constant coefficients reduces to calculation of the matrix $e^{tA} = e^{At}$. The explicit form of e^{At} is given in Secs. 19.5 and 20.6 for the case where the eigenvalues of the matrix are all distinct. We now use the Jordan normal form to find e^{At} in the case of multiple eigenvalues.

25.1. Calculation of e^{At} where A is a Jordan block.   One way of calculating e^{At}, where A is a Jordan block

$$\begin{pmatrix} \lambda & 1 & & \\ & \lambda & \cdot & \\ & & \cdot & \cdot & 1 \\ & & & \cdot & \cdot \\ & & & & \lambda \end{pmatrix} : \mathbf{R}^n \to \mathbf{R}^n,$$

was indicated in Sec. 14.9 (it will be recalled that the differentiation operator in the space of quasi-polynomials $e^{\lambda t} p_{<n}(t)$ has the matrix A in the basis $e_k = t^k e^{\lambda t}/k!$, $0 \leqslant k < n$). In fact, according to Taylor's formula, the matrix $H^s = e^{As}$ is the matrix of the shift operator $f(t) \mapsto f(s + t)$ in the indicated basis. Thus

$$e^{\lambda(t+s)} \frac{(t + s)^k}{k!} = \sum_l h_{kl}(s) \mathbf{e}_l,$$

where the elements $h_{kl}(s)$ of the matrix H^s are found by using the binomial theorem and turn out to be quasi-polynomials in s with exponents λ of degree less than n.

Another way of calculating e^{At} is based on the following

LEMMA.   *If the linear operators A, $B : \mathbf{R}^n \to \mathbf{R}^n$ commute, so that $AB = BA$, then $e^{A+B} = e^A e^B$.*

Proof.   Compare the formal series

$$e^A e^B = \left(E + A + \frac{A^2}{2!} + \cdots \right) \left(E + B + \frac{B^2}{2!} + \cdots \right)$$

$$= E + (A + B) + \tfrac{1}{2}(A^2 + 2AB + B^2) + \cdots,$$

$$e^{A+B} = E + (A + B) + \tfrac{1}{2}(A + B)^2 + \cdots$$

$$= E(A + B) + \tfrac{1}{2}(A^2 + AB + BA + B^2) + \cdots.$$

The series coincide if $AB = BA$, since $e^{x+y} = e^x e^y$ for $x, y \in \mathbf{R}$. But then $e^A e^B = e^{A+B}$, since the series are absolutely convergent. ∎

Suppose we represent A in the form

$$A = \lambda E + \Delta,$$

where

$$\Delta = \begin{pmatrix} 0 & 1 & & \\ & 0 & \cdot & \\ & & \cdot & \cdot \\ & & & \cdot & 1 \\ & & & & 0 \end{pmatrix}$$

is a nilpotent Jordan block. Since λE commutes with any operator, we have

$$e^{At} = e^{t(\lambda E + \Delta)} = e^{\lambda t} e^{\Delta t}.$$

THEOREM.   *The matrices $e^{\Delta t}$ and e^{At} are given by*

$$e^{\Delta t} = \begin{pmatrix} 1 & t & t^2/2 & \cdots & t^{n-1}/(n-1)! \\ & 1 & t & \cdots & \vdots \\ & & 1 & \cdot & t^2/2 \\ & & & \cdot & t \\ & & & & 1 \end{pmatrix},$$

$$e^{At} = \begin{pmatrix} e^{\lambda t} & te^{\lambda t} & \cdots & t^{n-1}e^{\lambda t}/(n-1)! \\ & e^{\lambda t} & \cdot & \vdots \\ & & \cdot & te^{\lambda t} \\ & & & e^{\lambda t} \end{pmatrix}. \tag{1}$$

Proof.   Since Δ operates on the basis $\mathbf{e}_1, \ldots, \mathbf{e}_n$ like a shift $0 \hookleftarrow e_1 \hookleftarrow e_2 \hookleftarrow \cdots \hookleftarrow e_3$, Δ^k acts like a shift by k places and has the matrix

$$\begin{pmatrix} 0 & \cdots & 1 & \\ & & & \cdot \\ & & & 1 \\ & & & \vdots \\ & & & 0 \end{pmatrix}.$$

But $e^{At} = e^{\lambda t} e^{\Delta t}$, where

$$e^{\Delta t} = E + \Delta t + \frac{\Delta^2 t^2}{2} + \cdots + \frac{\Delta^{n-1} t^{n-1}}{(n-1)!} \qquad (\Delta^n = 0).$$

The calculations go through without change in the complex case $(\lambda \in \mathbf{C}, A \colon \mathbf{C}^n \to \mathbf{C}^n)$. ∎

25.2. Implications.   Formula (1) immediately implies

COROLLARY 1.   *Let $A: \mathbf{C}^n \to \mathbf{C}^n$ be a linear operator with eigenvalues $\lambda_1, \ldots, \lambda_k$ of multiplicity $v_1, \ldots, v_k$ respectively. Then every element of the matrix of the operator e^{At}, $t \in \mathbf{R}$ (in any fixed basis) is a sum of quasi-polynomials in t, where the lth quasi-polynomial has exponent λ_l and is of degree less than v_l.*

Proof.   Consider the matrix of the operator e^{At} in the basis in which the matrix of A has Jordan form. The theorem then follows from (1), since the elements of the matrix of e^{At} in any other basis are linear combinations (with constant coefficients) of the elements of the matrix of e^{At} in the indicated basis. ∎

COROLLARY 2.   *Let $\boldsymbol{\varphi}$ be a solution of the differential equation*

$$\dot{\mathbf{x}} = A\mathbf{x}, \qquad \mathbf{x} \in \mathbf{C}^n, \qquad A: \mathbf{C}^n \to \mathbf{C}^n.$$

Then every component φ_j of the vector $\boldsymbol{\varphi}$ (in any fixed basis) is a sum of quasi-polynomials in t, where the lth quasi-polynomial p_{jl} has exponent λ_l and is of degree less than v_l:

$$\varphi_j(t) = \sum_{l=1}^{n} e^{\lambda_l t}\, p_{jl}(t).$$

Proof.   Merely note that $\boldsymbol{\varphi}(t) = e^{At}\boldsymbol{\varphi}(0)$. ∎

COROLLARY 3.   *Let $A: \mathbf{R}^n \to \mathbf{R}^n$ be a linear operator with real eigenvalues λ_l of multiplicity v_l $(1 \leqslant l \leqslant k)$ and complex eigenvalues $\alpha_l \pm i\beta_l$ of multiplicity μ_l $(1 \leqslant l \leqslant m)$. Then every element of the matrix of e^{At} and every component of the solution of the equation $\dot{\mathbf{x}} = A\mathbf{x}, \mathbf{x} \in \mathbf{R}^n$ is a sum of complex quasi-polynomials with exponents $\lambda_l, \alpha_l \pm i\omega_l$, where the degree of the quasi-polynomial with exponent λ_l is less than v_l and that of the quasi-polynomial with exponent $\alpha_l \pm i\omega_l$ is less than μ_l.*

Proof.   An immediate consequence of Corollaries 1 and 2. ∎

The sum figuring in Corollary 3 can also be written in the less convenient form

$$\varphi_j(t) = \sum_{l=1}^{k} e^{\lambda_l t}\, p_{jl}(t) + \sum_{l=1}^{m} e^{\alpha_l t}[q_{jl}(t) \cos \omega_l t + r_{jl}(t) \sin \omega_l t],$$

where p_{jl}, q_{jl}, r_{jl} are polynomials with *real* coefficients of degree less than v_l, μ_l, μ_l respectively. This representation follows from the fact that

$$\operatorname{Re} z e^{\lambda t} = \operatorname{Re} e^{\alpha t}(x + iy)(\cos \omega t + i \sin \omega t) = e^{\alpha t}(x \cos \omega t - y \sin \omega t)$$

if $z = x + iy, \lambda = \alpha + i\omega$. Moreover, it is clear from these formulas that if the real parts of all the eigenvalues are negative, then all the solutions approach 0 as $t \to +\infty$ (as must be the case, according to Secs. 22 and 23).

25.3. Applications to systems of higher-order equations.   Writing a system of higher-order equations as a system of first-order equations, we reduce the problem to the problem considered above, which in turn can be solved by reducing the matrix to Jordan form. In practice, however, it is often more convenient to proceed differently. First of all, we note that the eigenvalues of the equivalent first-order system can be found without writing down the matrix of the system. In fact, for every eigenvalue λ we have an eigenvector and hence a solution $\boldsymbol{\varphi}(t) = e^{\lambda t}\boldsymbol{\varphi}(0)$ of the equivalent first-order system. But then the original system has a solution of the form $\boldsymbol{\psi}(t) = e^{\lambda t}\boldsymbol{\psi}(0)$. Thus, substituting $\boldsymbol{\psi} = e^{\lambda t}\boldsymbol{\xi}$ into the original system, we see that the system has a (nonzero) solution of the given form if and only if λ satisfies a certain algebraic equation, from which the eigenvalues λ_l can be determined. We can then look for the solutions themselves in the form of sums of quasi-polynomials with exponents λ_l and undetermined coefficients.

Example 1. Let

$$x^{(iv)} = x. \tag{2}$$

Substituting $x = e^{\lambda t}\xi$ into (2), we get $\lambda^4 e^{\lambda t}\xi = e^{\lambda t}\xi$, $\lambda^4 = 1$, $\lambda_{1,2,3,4} = 1, -1, i, -i$. Thus every solution of (2) is of the form

$$x = C_1 e^t + C_2 e^{-t} + C_3 \cos t + C_4 \sin t.$$

Example 2. Let

$$\begin{cases} \dot{x}_1 = x_2, \\ \dot{x}_2 = x_1. \end{cases} \tag{3}$$

Substituting $\mathbf{x} = e^{\lambda t}\boldsymbol{\xi}$ into (3), we get $\lambda^2 \xi_1 = \xi_2$, $\lambda^2 \xi_2 = \xi_1$. This system of linear equations in ξ_1, ξ_2 has a nontrivial solution if and only if $\lambda^4 = 1$. Hence every solution of (3) is of the form

$$x_1 = C_1 e^t + C_2 e^{-t} + C_3 \cos t + C_4 \sin t,$$
$$x_2 = D_1 e^t + D_2 e^{-t} + D_3 \cos t + D_4 \sin t,$$

which gives

$$D_1 = C_1, \qquad D_2 = C_2, \qquad D_3 = -C_3, \qquad D_4 = -C_4$$

after substitution into (3).

Example 3. Let

$$x^{(iv)} - 2\ddot{x} + x = 0. \tag{4}$$

Substituting $x = e^{\lambda t}\xi$ into (4), we get

$$\lambda^4 - 2\lambda^2 + 1 = 0, \qquad \lambda^2 = 1, \qquad \lambda_{1,2,3,4} = 1, 1, -1, -1.$$

Thus every solution of (4) is of the form

$$x = (C_1 t + C_2)e^t + (C_3 t + C_4)e^{-t}.$$

Problem 1. Find the Jordan normal form of the fourth-order matrix corresponding to equation (4).

25.4. The case of a single equation of order n. In general, the multiplicity of the eigenvalues does not determine the sizes of the Jordan blocks. The situation becomes simpler if we are dealing with the linear operator A corresponding to a single differential equation of order n:

$$x^{(n)} = a_1 x^{(n-1)} + \cdots + a_n x, \qquad a_k \in \mathbf{C}. \tag{5}$$

Then Corollary 2 implies

COROLLARY 4.   *Every solution of equation* (5) *is of the form*

$$\varphi(t) = \sum_{l=1}^{k} e^{\lambda_l t} p_l(t), \tag{6}$$

where $\lambda_1, \ldots, \lambda_k$ *are the roots of the characteristic equation*

$$\lambda^n = a_1 \lambda^{n-1} + \cdots + a_n \tag{7}$$

of multiplicity $v_1, \ldots, v_l$ *respectively, and each* p_l *is a polynomial of degree less than* v_l.

Proof.   Equation (5) has a solution of the form $e^{\lambda t} \xi$ if and only if λ is a root of equation (7). ∎

Turning to the equivalent system of first-order equations

$$\dot{\mathbf{x}} = A\mathbf{x}, \quad A = \begin{pmatrix} 0 & 1 & & & \\ & 0 & 1 & & \\ & & \cdot & \cdot & \\ & & & \cdot & 1 \\ & & & \cdot & \cdot \\ a_n & & \cdots & & a_1 \end{pmatrix}. \tag{8}$$

we get

COROLLARY 5.   *If the operator* $A: \mathbf{C}^n \to \mathbf{C}^n$ *has a matrix of the form* (8), *then to every eigenvalue* λ *of* A *there corresponds precisely one Jordan block of size equal to the multiplicity of* λ.

Proof.   According to (6), there is a single eigendirection corresponding to every λ. In fact, let ξ be an eigenvector of the operator A. Then the first component $e^{\lambda t} \xi_0$ of the vector $e^{\lambda t} \xi$ is one of the solutions of (6). But then the remaining components are derivatives: $\xi_k = \lambda^k \xi_0$. Hence λ uniquely determines the direction of ξ. To complete the proof, we note that each Jordan block has its own eigendirection. ∎

Problem 1. Is every linear combination of quasi-polynomials (6) a solution of equation (5) ?

25.5. Recurrent sequences. Our study of the exponential e^{At} with a continuous argument t can easily be carried over to the case of the exponential A^n with the discrete argument n. In particular, we can investigate any recurrent sequence defined by a relation of the form

$$x_n = a_1 x_{n-1} + \cdots + a_k x_{n-k} \tag{9}$$

(for example, the sequence 0, 1, 2, 5, 12, 29, ... specified by the relation $x_n = 2x_{n-1} + x_{n-2}$ and the initial condition $x_0 = 0$, $x_1 = 1$).

COROLLARY 6. *The nth term of the recurrent sequence defined by* (9) *depends on n like a sum*

$$x_n = \sum_{l=1}^{k} \lambda^n p_l(n)$$

of quasi-polynomials in n, where $\lambda_1, \ldots, \lambda_k$ are the eigenvalues of the matrix corresponding to the sequence, of multiplicity $v_1, \ldots, v_k$ respectively, and each p_l is a polynomial of degree less than v_l.

Proof. First we note that the matrix in question is the matrix of the operator $A : \mathbf{R}^k \to \mathbf{R}^k$ carrying the section $\xi_{n-1} = (x_{n-k}, \ldots, x_{n-1})$ of length k of our sequence into the next section $\xi_n = (x_{n-k+1}, \ldots, x_n)$ of length k:

$$A\xi_{n-1} = \begin{pmatrix} 0 & 1 & & & \\ & 0 & 1 & & \\ & & \ddots & \ddots & \\ & & & 0 & 1 \\ a_k & \cdots & a_2 & a_1 \end{pmatrix} \begin{pmatrix} x_{n-k} \\ \vdots \\ \vdots \\ x_{n-1} \end{pmatrix} = \begin{pmatrix} x_{n-k+1} \\ \vdots \\ \vdots \\ x_n \end{pmatrix} = \xi_n.$$

It is important to note that the operator A does not depend on n. Hence x_n is one of the components of the vector $A^n \xi$, where ξ is a constant vector and the matrix of A is of the form (5). We now apply Corollary 5, reducing the matrix of A to Jordan form. ∎

In making the calculations, there is no need either to write down the matrix or reduce it to normal form. In fact, any eigenvalue of the operator A corresponds to a solution of equation (9) of the form $x_n = \lambda^n$. Substituting $x_n = \lambda^n$ into (9), we find that λ satisfies the equation

$$\lambda^k = a_1 \lambda^{k-1} + \cdots + a_k,$$

which, as is easily verified, is just the characteristic equation of the operator A.

Example 1. For the sequence 0, 1, 2, 5, 12, 29, ... corresponding to the relation

$$x_n = 2x_{n-1} + x_{n-2}, \tag{10}$$

we have $\lambda^2 = 2\lambda + 1$, $\lambda_{1,2} = 1 \pm \sqrt{2}$. Hence the sequences

$$x_n = (1 + \sqrt{2})^n, \quad x_n = (1 - \sqrt{2})^n$$

both satisfy (10), and so do all linear combinations

$$x_n = c_1(1 + \sqrt{2})^n + c_2(1 - \sqrt{2})^n$$

of these sequences (and only such linear combinations). Among these combinations, it is easy to find the one for which $x_0 = 0$, $x_1 = 1$. In fact, solving the equations

$$c_1 + c_2 = 0, \qquad \sqrt{2}(c_1 - c_2) = 1,$$

we find that

$$x_n = \frac{(1 + \sqrt{2})^n}{2\sqrt{2}} - \frac{(1 - \sqrt{2})^n}{2\sqrt{2}}.$$

Comment. As $n \to \infty$, the first term increases exponentially, while the second term decreases exponentially. Therefore

$$x_n \approx \frac{(1 + \sqrt{2})^n}{2\sqrt{2}}$$

for large n, and in particular $x_{n+1}/x_n \approx 1 + \sqrt{2}$. This gives us very good approximations to $\sqrt{2}$:

$$\sqrt{2} \approx \frac{x_{n+1} - x_n}{x_n}.$$

Choosing $x_n = 0, 1, 2, 5, 12, 29, \ldots$, we get

$$\sqrt{2} \approx \frac{1 - 0}{1} = 1, \qquad \sqrt{2} \approx \frac{5 - 2}{2} = 1.5,$$

$$\sqrt{2} \approx \frac{12 - 5}{5} = 1.4, \qquad \sqrt{2} \approx \frac{29 - 12}{12} = 1.417 \ldots$$

These are the same approximations used to calculate $\sqrt{2}$ in ancient times, and can be obtained by expanding $\sqrt{2}$ in a continuous fraction. Moreover $(x_{n-1} - x_n)/x_n$ is the best of all rational approximations to $\sqrt{2}$ with denominators not exceeding x_n.

25.6. Small oscillations. In Sec. 25.4 we considered the case where to each root of the characteristic equation, regardless of its multiplicity, there corresponds a single eigenvector, namely the case of a single equation of order n. We now consider the case (which, in a certain sense, is the opposite of the one just cited) where each root has a number of eigenvectors equal to its multiplicity. This is the case of small oscillations of a conservative mechanical system.

Let U be a quadratic form in the *Euclidean space* $\mathbf{R}^n$ given by a symmetric operator A, i.e., let

$$U(x) = \tfrac{1}{2}(A\mathbf{x}, \mathbf{x}), \qquad \mathbf{x} \in \mathbf{R}^n, \quad A: \mathbf{R}^n \to \mathbf{R}^n, \quad A' = A,$$

where A' denotes the transpose of A, and consider the differential equation†

$$\ddot{\mathbf{x}} = -\operatorname{grad} U, \tag{11}$$

---

† The vector field grad U is defined by the condition that $dU(\xi) = (\operatorname{grad} U, \xi)$ for every vector $\xi \in T\mathbf{R}^n_x$, where $(\cdot, \cdot)$ denotes the Euclidean scalar product. In (orthonormal) rectangular coordinates, the field grad U has components $\partial U/\partial x_1, \ldots, \partial U/\partial x_n$.

thinking of U as the potential energy. In investigating (11), it is useful to imagine a bead sliding down the graph of the potential energy (see Sec. 12). Equation (11) can also be written in the form

$$\ddot{\mathbf{x}} = -A\mathbf{x},$$

or as a system of n linear second-order equations in the coordinates of $\mathbf{x}$. Following our general rule, we look for a solution of the form $\boldsymbol{\varphi} = e^{\lambda t}\boldsymbol{\xi}$. This gives

$$\lambda^2 e^{\lambda t}\boldsymbol{\xi} = -Ae^{\lambda t}\boldsymbol{\xi}, \qquad (A + \lambda^2 E)\boldsymbol{\xi} = 0, \qquad \det(A + \lambda^2 E) = 0.$$

It follows that λ^2 has n real values (why?), and correspondingly that λ has $2n$ real or purely imaginary values. If these values are all different, then every solution of (11) is a linear combination of exponentials. If there are multiple roots, we encounter the problem of Jordan blocks.

THEOREM. *If the quadratic form U is nondegenerate, then each eigenvalue λ has a number of linearly independent eigenvectors equal to its multiplicity. Correspondingly, every solution of equation* (11) *can be written as a sum of exponentials*:[†]

$$\boldsymbol{\varphi}(t) = \sum_{k=1}^{2n} e^{\lambda_k t}\boldsymbol{\xi}_k, \qquad \boldsymbol{\xi}_k \in \mathbf{C}^n. \tag{12}$$

Proof. The form U can be reduced to *principal axes* by making an orthogonal transformation, i.e., there exists an orthonormal basis $\mathbf{e}_1, \ldots, \mathbf{e}_n$ in which U becomes

$$U(x) = \frac{1}{2}\sum_{k=1}^{n} a_k^2 x_k^2, \qquad \mathbf{x} = x_1\mathbf{e}_1 + \cdots + x_n\mathbf{e}_n.$$

Since U is nondegenerate, none of the numbers a_k vanishes. In the indicated coordinates, equation (11) becomes

$$\ddot{x}_1 = -a_1 x_1, \qquad \ddot{x}_2 = -a_2 x_2, \ldots, \qquad \ddot{x}_n = -a_n x_n,$$

whether or not there are multiple roots.[‡] Thus our system decomposes into the direct product of n "pendulum equations." Each of these equations ($\ddot{x} = -ax$) can be solved at once. In fact, if $a > 0$, then $a = \omega^2$ and

$$x = C_1\cos \omega t + C_2\sin \omega t,$$

[†] It is interesting to note that Lagrange, who first investigated the equation of small oscillations (11), initially made a mistake, thinking that "secular" terms of the form $te^{\lambda t}$ (or $t \sin \omega t$ in the real case) were required in the case of multiple roots, as in the earlier part of this section.

[‡] Note that we have made essential use of the orthonormality of the basis $\mathbf{e}_1, \ldots, \mathbf{e}_n$. If the basis were not orthonormal, the components of the vector grad $\frac{1}{2}\sum a_k x_k^2$ *would not equal* $a_k x_k$.

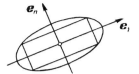

Fig. 173   Level curve of the potential energy and directions of the characteristic oscillations.

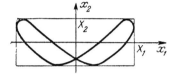

Fig. 174   One of the Lissajous figures with $\omega_1 = 1$, $\omega_2 = 2$.

while if $a < 0$, then $a = -\alpha^2$ and

$$x = C_1 \cosh \alpha t + C_2 \sinh \alpha t = D_1 e^{\alpha t} + D_2 e^{-\alpha t}.$$

In particular, these formulas immediately imply (12). ∎

If the form U is positive definite, then the a_k are all positive and the point x executes n independent oscillations (called "normal modes") along the n mutually perpendicular directions $\mathbf{e}_1, \ldots, \mathbf{e}_n$ (Fig. 173). The numbers ω_k, which satisfy the equation $\det(A - \omega^2 E) = 0$, are called the *characteristic* (or *natural*) *frequencies*. The trajectory of the point $\mathbf{x} = \boldsymbol{\varphi}(t)$ in $\mathbf{R}^n$, where $\boldsymbol{\varphi}$ is a solution of (11), lies in the parallelepiped $|x_k| \leq X_k$, $1 \leq k \leq n$, where X_k is the amplitude of the kth characteristic oscillation. The parallelepiped reduces to a rectangle if $n = 2$.

If the frequencies ω_1 and ω_2 are commensurable, the trajectory is a closed curve, called a *Lissajous figure* (Fig. 174). However, if ω_1 and ω_2 are incommensurable, the trajectory fills the whole rectangle densely.

Problem 1.   Draw Lissajous figures for $\omega_1 = 1$, $\omega_2 = 3$ and $\omega_1 = 3$, $\omega_1 = 1$.

Problem 2.   Prove that one of the Lissajous figures with $\omega_2 = n\omega_1$ is the graph of the polynomial

$$T_n(x) = \cos(n \arccos x)$$

of degree n, called a *Chebyshev polynomial*.

Problem 3.   What do the trajectories $\mathbf{x} = \boldsymbol{\varphi}(t)$ look like if $U = x_1^2 - x_2^2$?

Problem 4.   For which U is the equilibrium position $x = \dot{x} = 0$ of equation
(11) stable
(a) In Lyapunov's sense; (b) asymptotically?

26. More on Quasi-Polynomials

In solving linear equations with constant coefficients, we have repeatedly
encountered quasi-polynomials. We now explain the reason for this
phenomenon, and give some further applications of quasi-polynomials.

26.1. The space of infinitely differentiable functions.   Let F be the
set of all complex-valued infinitely differentiable functions defined on the
real line **R**. The set F has the natural structure of a complex linear space,
since

$$f_1, f_2 \in F, \qquad c_1, c_2 \in \mathbf{C}$$

obviously implies $c_1 f_1 + c_2 f_2 \in F$.

Definition.   The functions $f_1, \ldots, f_n \in F$ are said to be *linearly independent* if
they are linearly independent regarded as vectors of the linear space F, i.e.,
if

$$c_1 f_1 + \cdots + c_n f_n = 0 \qquad (c_1, \ldots, c_n \in \mathbf{C})$$

implies $c_1 = \cdots = c_n = 0$.

Problem 1. For what values of α and β are the functions sin αt and sin βt linearly dependent?

Problem 2. Prove that the functions $e^{\lambda_1 t}, \ldots, e^{\lambda_n t}$ are linearly independent if the numbers
λ_k are distinct.

Hint. This follows from the existence of a linear equation of order n with solutions
$e^{\lambda_1 t}, \ldots, e^{\lambda_n t}$ (see Sec. 26.2).

The space F contains all quasi-polynomials

$$f(t) = e^{\lambda t} \sum_{k=0}^{\nu - 1} c_k t^k$$

with exponent λ and, more generally, all finite sums

$$f(t) = \sum_{l=1}^{k} e^{\lambda_l t} \sum_{m=0}^{\nu_l - 1} c_{lm} t^m, \qquad \lambda_i \neq \lambda_j \tag{1}$$

of quasi-polynomials with different exponents.

Problem 3. Prove that every function which can be represented by a sum of the form (1)
has a unique representation of this form. In other words, prove that *if the sum* (1) *vanishes
identically, then every coefficient c_{lm} equals* 0.

Hint. For one possible solution see the corollary below.

26.2. The space of solutions of a linear equation.

THEOREM.   *The set X of all solutions of the linear equation*

$$x^{(n)} + a_1 x^{(n-1)} + \cdots + a_n x = 0 \tag{2}$$

is an n-dimensional linear subspace of F.

Proof.   Consider the operator $D: F \to F$ carrying every function into its derivative. The operator D is linear:

$$D(c_1 f_1 + c_2 f_2) = c_1 D f_1 + c_2 D f_2.$$

Let

$$A = a(D) = D^n + a_1 D^{n-1} + \cdots + a_n E$$

be a polynomial in the operator D. Then A is a linear operator $A: F \to F$. The solutions of equation (2) are just the elements of the kernel of A, † so that $X = \text{Ker } A$. But the kernel of a linear operator is a linear space, and hence X is a linear space.

Next we show that X is isomorphic to $\mathbf{C}^n$. Given any $\varphi \in X$, we associate with φ a set of n numbers, namely the values of the function φ and its first $n - 1$ derivatives at the point $t = 0$:

$$\varphi_0 = (\varphi(0), (D\varphi)(0), \ldots, (D^{n-1}\varphi)(0)).$$

This gives a mapping

$$B: X \to \mathbf{C}^n, \qquad B\varphi = \varphi_0,$$

which is clearly linear. The image of B is the whole space $\mathbf{C}^n$, since by the existence theorem, there exists a solution $\varphi \in X$ with any given initial conditions φ_0. Moreover, the kernel of the mapping B consists of the single element 0, since by the uniqueness theorem, the initial conditions $\varphi_0 = 0$ uniquely determine the solution $(\varphi \equiv 0)$. Therefore B is an isomorphism. ∎

COROLLARY.   *Let $\lambda_1, \ldots, \lambda_k$ be the roots of the characteristic equation $a(\lambda) = 0$ of the differential equation (2), of multiplicity $v_1, \ldots, v_k$ respectively. Then every solution of equation (2) has a unique representation of the form (1) and every sum of quasi-polynomials of the form (1) satisfies equation (2).*

Proof.   Formula (1) gives a mapping $\Phi: \mathbf{C}^n \to F$, associating a function f with every set of n coefficients c_{lm}. The mapping Φ is linear. Moreover the image of Φ contains the space X of all solutions of equation (2), since according to Sec. 25.4, every solution of equation (2) can be written in the

† We already know that all the solutions of equation (2) are infinitely differentiable, i.e., belong to F (see Sec. 25.4).

form (1). By the above theorem, the dimension of X equals n. But a linear mapping of the space $\mathbf{C}^n$ onto a space X of the same dimension is an isomorphism. Therefore Φ establishes an isomorphism between $\mathbf{C}^n$ and X. ∎

26.3. Invariance under shifts.

THEOREM. *The space X of solutions of the differential equation* (2) *is invariant under shifts carrying the function $\varphi(t)$ into $\varphi(t + s)$.*

Proof. The shift of a solution is a solution, as in the case of any autonomous equation (Sec. 10.1). ∎

The following are all examples of shift-invariant subspaces of the space F:

Example 1. The one-dimensional space $\{ce^{\lambda t}\}$.

Example 2. The space of quasi-polynomials $\{e^{\lambda t}p_{<n}(t)\}$ of dimension n.

Example 3. The plane $\{c_1 \cos \omega t + c_2 \sin \omega t\}$.

Example 4. The space $\{p_{<n}(t) \cos \omega t + q_{<n}(t) \sin \omega t\}$ of dimension $2n$.

It can be shown that every finite-dimensional shift-invariant subspace of the space F is the space of solutions of some differential equation (2). In other words, such a subspace always decomposes into a direct sum of spaces of quasi-polynomials. This explains the significance of quasi-polynomials in the theory of linear differential equations with constant coefficients.

If an equation is invariant under some group of transformations, then the space of functions invariant under the group will play an important role in solving the equation. This is how various special functions arise in mathematics. For example, there is a connection between the group of rotations of the sphere and the finite-dimensional spaces of functions on the sphere ("spherical functions") which are invariant under rotations.

Problem 1. Find all finite-dimensional subspaces of the space of smooth functions on the circle which are invariant under rotations of the circle.

26.4. Historical remark.

The theory of linear differential equations with constant coefficients was created by Euler and Lagrange before the discovery of the Jordan normal form of a matrix. They reasoned as follows: Let λ_1 and λ_2 be two roots of the characteristic equation. The solutions $e^{\lambda_1 t}$ and $e^{\lambda_2 t}$ corresponding to these roots span a two-dimensional plane $\{c_1 e^{\lambda_1 t} + c_2 e^{\lambda_2 t}\}$ in the space F (Fig. 175). Suppose the equation changes in such a way that λ_2 approaches λ_1. Then $e^{\lambda_2 t}$ approaches $e^{\lambda_1 t}$ and the plane degenerates into a line for $\lambda_2 = \lambda_1$. The question now arises of whether the plane has a limiting position as $\lambda_2 \to \lambda_1$. If $\lambda_2 \neq \lambda_1$, we can choose $e^{\lambda_1 t}$, $e^{\lambda_2 t} - e^{\lambda_1 t}$ rather than $e^{\lambda_1 t}$, $e^{\lambda_2 t}$ as the basis. But

$$e^{\lambda_2 t} - e^{\lambda_1 t} \approx (\lambda_2 - \lambda_1)te^{\lambda_1 t},$$

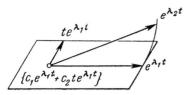

Fig. 175   Limiting position of the plane spanned by $e^{\lambda_1 t}$ and $e^{\lambda_2 t}$ in the space F.

and hence, as $\lambda_2 \to \lambda_1$, the plane spanned by $e^{\lambda_1 t}$ and $e^{\lambda_2 t} - e^{\lambda_1 t}$, or equivalently the plane spanned by $e^{\lambda_1 t}$ and $(e^{\lambda_2 t} - e^{\lambda_1 t})/(\lambda_2 - \lambda_1)$, goes into the limiting plane spanned by $e^{\lambda_1 t}$ and $te^{\lambda_1 t}$. Therefore it is natural to expect that the solutions of the limiting equation (with the double root $\lambda_2 = \lambda_1$) will lie in the limiting plane $\{c_1 e^{\lambda_1 t} + c_2 te^{\lambda_1 t}\}$, where the fact that $c_1 e^{\lambda_1 t} + c_2 te^{\lambda_1 t}$ is a solution of the orginal differential equation can be verified by direct substitution. The same reasoning explains the appearance of the solutions $t^k e^{\lambda t}$ $(k < v)$ in the case of a v-fold root.

The above argument can easily be made perfectly rigorous (for example, with the help of the theorem on differentiable dependence of the solutions on a parameter).

26.5. Nonhomogeneous equations.   Given a linear operator $A : L_1 \to L_2$, by a *solution* of the nonhomogeneous equation

$$Ax = f$$

with right-hand side f is meant any preimage $x \in L_1$ of the element $f \in L_2$ (Fig. 176). Every solution of the nonhomogeneous equation is the sum of a particular solution x_1 and the general solution of the homogeneous equation $Ax = 0$:

$$A^{-1}f = x_1 + \text{Ker } A.$$

The nonhomogeneous equation is solvable if and only if f belongs to the linear space $\text{Im } A = A(L_1) \subset L_2$.

In particular, consider the differential equation

$$x^{(n)} + a_1 x^{(n-1)} + \cdots + a_n x = f(t) \tag{3}$$

(*a nonhomogeneous linear equation of order n with constant coefficients*).

THEOREM.   *If the right-hand side $f(t)$ of equation (3) is a sum of quasipolynomials, then so is every solution of equation (3).*

Let

$$\mathbf{Q}^m = \{e^{\lambda t} p_{<m}(t)\}$$

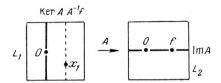

Fig. 176   Kernel and image of an operator A.

be the space of all quasi-polynomials of degree less than m and exponent λ. The linear operator D (carrying every function into its derivative) carries $\mathbf{Q}^m$ into itself, and hence the operator

$$A = a(D) = D^n + a_1 D^{n-1} + \cdots + a_n E : \mathbf{Q}^m \to \mathbf{Q}^m$$

is also a linear operator from $\mathbf{Q}^m$ into itself. We can now write equation (3) in the form $Ax = f$, and to investigate the solvability of (3), we must find the image Im $A = A(\mathbf{Q}^m)$ of the mapping A.

LEMMA 1.   *Suppose λ is not a root of the characteristic equation, so that $a(\lambda) \neq 0$. Then $A : \mathbf{Q}^m \to \mathbf{Q}^m$ is an isomorphism.*

Proof.   The matrix of the operator $D : \mathbf{Q}^m \to \mathbf{Q}^m$ in a suitable basis is the Jordan block with λ on the main diagonal. In the same basis, the operator A has a triangular matrix with $a(\lambda)$ on the main diagonal. Hence

$$\det A = [a(\lambda)]^m \neq 0,$$

and A is an isomorphism.   ∎

COROLLARY 1.   *Suppose λ is not a root of the characteristic equation, and suppose equation (3) has a quasi-polynomial of degree less than m and exponent λ as its right-hand side. Then equation (3) has a particular solution which is also a quasi-polynomial of degree less than m and exponent λ.*

Proof.   An immediate consequence of Lemma 1.   ∎

LEMMA 2.   *Suppose λ is a root of the characteristic equation of multiplicity ν, so that*

$$a(z) = (z - \lambda)^\nu b(z), \qquad b(\lambda) \neq 0.$$

Then

$$A\mathbf{Q}^m = \mathbf{Q}^{m-\nu}.$$

Proof.   Here

$$A = a(D) = (D - \lambda E)^\nu b(D),$$

where $b(D) : \mathbf{Q}^m \to \mathbf{Q}^m$ is an isomorphism, by Lemma 1. It remains to show that $(D - \lambda E)^\nu \mathbf{Q}^m = \mathbf{Q}^{m-\nu}$. But the matrix of the operator $D - \lambda E$ in the basis

$$\mathbf{e}_k = \frac{t^k}{k!}e^{\lambda t}, \qquad 0 \leqslant k < m$$

is a nilpotent Jordan block, i.e., $D - \lambda E$ acts on the basis like a shift:

$$0 \leftarrow\!\!\!\leftarrow \mathbf{e}_0 \leftarrow\!\!\!\leftarrow \mathbf{e}_1 \leftarrow\!\!\!\leftarrow \cdots \leftarrow\!\!\!\leftarrow \mathbf{e}_{m-1}.$$

Hence the operator $(D - \lambda E)^\nu$ acts like a shift by ν places and maps $\mathbf{Q}^m$ onto $\mathbf{Q}^{m-\nu}$. ∎

COROLLARY 2.   *Let λ be a root of multiplicity ν of the characteristic equation $a(\lambda) = 0$, and let $f \in \mathbf{Q}^k$ be a quasi-polynomial of degree less than k and exponent λ. Then equation (3) has a solution $\varphi \in \mathbf{Q}^{k+\nu}$ which is a quasi-polynomial of degree less than $k + \nu$ and exponent λ.*

Proof.   We need only set $m = k + \nu$ in Lemma 2. ∎

Proof of the theorem.   Let Σ be the set of all possible sums of quasi-polynomials. Then Σ is an infinite-dimensional subspace of the space F. By Corollary 2, the image $A(\Sigma)$ of the operator

$$A = a(D) : \Sigma \to \Sigma$$

contains all quasi-polynomials. Hence $A(\Sigma)$ coincides with Σ, being a linear space. Therefore equation (3) has a particular solution which is a sum of quasi-polynomials. It remains only to add the general solution of the homogeneous equation, which, according to Sec. 25.4, is itself a sum of quasi-polynomials. ∎

Remark 1.   If $f = e^{\lambda t}p_{<k}(t)$, *then equation (3) has a particular solution of the form $\varphi = t^\nu e^{\lambda t}q_{<k}(t)$.* In fact, there exists a particular solution in the form of a quasi-polynomial of degree less than $k + \nu$. But the terms of degree less than ν satisfy the homogeneous equation (see Sec. 25.4) and hence can be dropped.

Remark 2.   Suppose equation (3) is real. Then we can look for a solution in the form of a real quasi-polynomial if λ is real, and in the form

$$e^{\alpha t}[p(t)\cos \omega t + q(t)\sin \omega t]$$

if $\lambda = \alpha \pm i\omega$. Here the solution can contain a sine function even in the case where the right-hand side of (3) consists only of a cosine.

Problem 1. Find the form of the particular solution of each of the following equations:
a) $\ddot{x} \pm x = t^2$;   b) $\ddot{x} \pm x = e^{2t}$;   c) $\ddot{x} \pm x = te^{-t}$;   d) $\ddot{x} \pm x = t^3 \sin t$;
e) $\ddot{x} \pm x = te^t \cos t$;   f) $x \pm 2ix = t^2 e^t \sin t$;   d) $x^{(iv)} + 4x = t^2 e^t \cos t$.

26.6. The method of complex amplitudes. In the case of complex roots, it is usually simpler to carry out the calculations as follows. Let equation (3) be real, and represent $f(t)$ as the real part of a complex function:

$$f(t) = \operatorname{Re} F(t).$$

Let $\Phi(t)$ be a complex solution of the equation

$$a(D)\Phi(t) = F(t).$$

Then, taking real parts, we see that

$$a(D)\varphi(t) = f(t),$$

where $\varphi = \operatorname{Re} \Phi$ (since $a = \operatorname{Re} a$). Thus *to solve a nonhomogeneous linear equation with right-hand side $f(t)$, we need only regard $f(t)$ as the real part of a complex function $F(t)$, solve the equation with right-hand side $F(t)$, and take the real part of the solution.*

Example 1. Let

$$f(t) = \cos \omega t = \operatorname{Re} e^{i\omega t}.$$

The quasi-polynomial $F(t) = e^{i\omega t}$ is of degree 0, and hence we can look for a solution Φ of the form $Cte^{i\omega t}$, where C is a complex constant (called the *complex amplitude*) and v is the multiplicity of the root $i\omega$. Therefore

$$\varphi(t) = \operatorname{Re} (Ct^v e^{i\omega t}).$$

If $C = re^{i\theta}$, then $\varphi(t) = rt^v \cos(\omega t + \theta)$. Thus the complex amplitude C contains information about both the amplitude r and the phase θ of the real solution.

Example 2. Consider the behavior of a pendulum (or of any other oscillatory linear system, for example a weight on a spring or an oscillatory electric circuit) under the action of an external periodic force:

$$\ddot{x} + \omega^2 x = f(t), \qquad f(t) = \cos vt = \operatorname{Re} e^{ivt}$$

(Fig. 177). The characteristic equation $\lambda^2 + \omega^2 = 0$ has roots $\lambda = \pm i\omega$. If $v^2 \neq \omega^2$, we must look for a particular solution of the form $\Phi = Ce^{ivt}$. Substituting Φ into the differential equation, we get the quantity

$$C = \frac{1}{\omega^2 - v^2}, \tag{4}$$

which can be written in trigonometric form as

$$C = r(v)e^{i\theta(v)}. \tag{5}$$

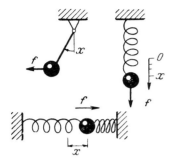

Fig. 177   An oscillatory system under the action of an external force $f(t) = \cos \nu t$.

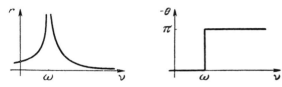

Fig. 178   The amplitude and phase of forced oscillations of a frictionless pendulum as a function of the frequency of the external force.

According to (4), the amplitude r and the phase θ have the values shown in Fig. 178.† The real part of Φ equals $r \cos(\nu t + \theta)$. Hence the general solution of the nonhomogeneous equation is of the form

$$x = r \cos(\nu t + \theta) + C_1 \cos(\omega t + \theta_1),$$

where C_1 and θ_1 are arbitrary constants.

Thus *the oscillations of a pendulum under the action of an external force consist of "forced oscillations"* $r \cos(\nu t + \theta)$ *with the frequency of the external force and "free oscillations" with the natural frequency* ω. The dependence of the amplitude r of the forced oscillations on the frequency of the external force has the characteristic resonance shape: The nearer the frequency of the external force to the natural frequency ω, the more the external force "rocks" the system. This phenomenon of resonance, observed when the frequency of the external force coincides with the natural frequency of the oscillatory system, is very important in the applications. For example, in all kinds of calculations involving engineering structures, care must be taken to see that the natural frequencies of the structure are not close to the frequencies of the external forces which will be experienced by the structure. Otherwise even a small force, acting over a long time interval, will be able to rock the structure and destroy it.

† The choice $\theta = -\pi$ (rather than $+\pi$) for $\nu > \omega$ is justified by Example 3 below.

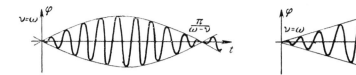

Fig. 179   The sum of two harmonics with neighboring frequencies (beats) and its limit in the case of resonance ("rocking").

The phase θ of the forced oscillations undergoes a jump of $-\pi$ as v passes through the resonance frequency ω. When v is near ω, "beats" are observed (Fig. 179), i.e., the amplitude of the pendulum alternately waxes (when the relation of the phases of the pendulum and the external force is such that the external force rocks the pendulum, communicating energy to it) and wanes (when the relation between the phases changes in such a way that the external force "brakes" the pendulum). The closer the frequencies v and ω, the more slowly the phase relation changes and the larger the period of the beats. As $v \to \omega$, the period of the beats approaches infinity. At resonance ($v = \omega$) the phase relation is constant and the forced oscillations can grow indefinitely. In fact, for $v = \omega$ we look for a particular solution of the form

$$x = \operatorname{Re}(Cte^{i\omega t}), \tag{6}$$

in accordance with the general rule. Substituting (6) into the differential equation, we get $C = 1/2i\omega$ and hence

$$x = \frac{t}{2\omega}\sin \omega t,$$

so that the forced oscillations grow without limit (Fig. 179).

Example 3.   Consider the pendulum with friction:

$$\ddot{x} + k\dot{x} + \omega^2 x = f(t).$$

The corresponding characteristic equation

$$\lambda^2 + k\lambda + \omega^2 = 0$$

has roots

$$\lambda_{1,2} = -\alpha \pm i\Omega, \qquad \alpha = -\frac{k}{2}, \qquad \Omega = \sqrt{\omega^2 - \frac{k^2}{4}}$$

(Fig. 180). Suppose the coefficient of friction k is positive and small ($k^2 < 4\omega^2$), and let the external force be oscillatory:

$$f(t) = \cos vt = \operatorname{Re} e^{ivt}.$$

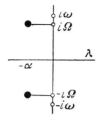

Fig. 180   Eigenvalues of the equation of the pendulum with friction.

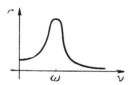 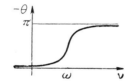

Fig. 181   The amplitude and phase of forced oscillations of a pendulum with friction as a function of the frequency of the external force.

If the coefficient of friction is different from 0, then iv cannot be a root of the characteristic equation (since $\lambda_{1,2}$ has a nonzero real part). Hence we should look for a solution of the form

$$x = \operatorname{Re} Ce^{ivt}. \tag{7}$$

Substitution of (7) into the differential equation gives

$$C = \frac{1}{\omega^2 - v^2 + ikv}. \tag{8}$$

Suppose we write C in the trigonometric form (5). Then, according to (7), the graphs of the amplitude r and phase θ of the forced oscillations, as functions of the frequency v of the external force, have the appearance shown in Fig. 181.

Adding the general solution $C_1 e^{-\alpha t}\cos(\Omega t + \theta_1)$ of the homogeneous equation to the particular solution, we get the general solution

$$x = r \cos(vt + \theta) + C_1 e^{-\alpha t}\cos(\Omega t + \theta_1)$$

of the nonhomogeneous equation. The second term on the right approaches 0 as $t \to +\infty$, leaving only the forced oscillations $x = r\cos(vt + \theta)$.

Comparing the behavior of the frictionless pendulum (Fig. 178) with its behavior for positive values of the coefficient of friction (Fig. 181), we find that *the effect of small friction on the resonance is such that the amplitude of the oscillations at resonance do not become infinite, but rather increase to a definite finite value which is inversely proportional to the coefficient of friction.* In fact, the function $r(v)$ expressing the dependence of the amplitude of the steady-state oscillations on the frequency of the external force has a sharply defined maximum near $v = \omega$ (Fig. 181), and it is clear from (8) that the height of this maximum increases like $1/k\omega$ as k decreases.

From a "physical" point of view, it is easy to predict the fact that the amplitude of the steady-state forced oscillations is finite, by simply calculating the energy balance. At large amplitudes, the energy loss due to friction is greater than the energy communicated to the pendulum by the external force, and hence the amplitude will decrease until a regime is established in which the energy loss due to friction equals the work done by the external force. The size of the amplitude of the steady-state oscillations increases in inverse proportion to the coefficient of friction k as $k \to 0$. The phase shift θ is always negative, i.e., *the forced oscillations always lag the external force.*

Problem 1. Prove that every solution of a nonhomogeneous linear system of equations with constant coefficients and a right-hand side

$$\mathbf{f} = \sum_l e^{\lambda_l t} \sum_k \mathbf{C}_{kl}\, t^k,$$

equal to a sum of quasi-polynomials with vector coefficients, is itself a sum of quasi-polynomials with vector coefficients.

Problem 2. Show that every solution of a nonhomogeneous linear recurrence relation

$$x_n - (a_1 x_{n-1} + \cdots + a_k x_{n-k}) = f(n)$$

with a right-hand side equal to a sum of quasi-polynomials is itself a sum of quasi-polynomials. Find a formula for the general term of the sequence $0, 2, 7, 18, 41, 88, \ldots$ $(x_n = 2x_{n-1} + n)$.

26.7. Application to the calculation of weakly nonlinear oscillations. In studying the dependence of the solution of an equation on parameters, we have to solve a nonhomogeneous linear equation, namely the "equation of variations" (see Sec. 9.5). In particular, if the "unperturbed" system is linear, the problem often reduces to the solution of a linear equation with a right-hand side which is a sum of exponentials (or trigonometric functions) or quasi-polynomials.

Problem 1. Find the dependence of the period of oscillations of a pendulum described by the equation $\ddot{x} = -\sin x$ on the amplitude A, assuming that A is small.

Ans. $T = 2\pi[1 + (A^2/16) + O(A^4)]$. For example, if the angle of deviation is $30°$, the period exceeds the period of small oscillations by 2 percent.

Solution. Consider the solution of the pendulum equation with initial condition $x(0) = A$, $\dot{x}(0) = 0$ as a function of A. This function is smooth, by the theorem on differentiable dependence on the initial conditions. Expanding the function in Taylor series in A near

$A = 0$, we get

$$x = Ax_1(t) + A^2x_2(t) + A^3x_3(t) + O(A^4),$$

so that

$$\dot{x} = A\dot{x}_1 + A^2\dot{x}_2 + A^3\dot{x}_3 + O(A^4),$$
$$\ddot{x} = A\ddot{x}_1 + A^2\ddot{x}_2 + A^3\ddot{x}_3 + O(A^4),$$
$$\sin x = Ax_1 + A^2x_2 + A^3(x_3 - \tfrac{1}{6}x_1^3) + O(A^4).$$

The equation $\ddot{x} = -\sin x$ holds for every A, and hence x_1, x_2, x_3 satisfy the equations

$$\ddot{x}_1 = -x_1, \qquad \ddot{x}_2 = -x_2, \qquad \ddot{x}_3 = -x_3 + \tfrac{1}{6}x_1^3. \tag{9}$$

The initial condition $x(0) = A$, $\dot{x}(0) = 0$ also holds for every A, and hence the equations (9) satisfy the following initial conditions:

$$x_1(0) = 1, \qquad x_2(0) = x_3(0) = \dot{x}_1(0) = \dot{x}_2(0) = \dot{x}_3(0) = 0. \tag{10}$$

Solving the first two equations (9) subject to the conditions (10), we get

$$x_1 = \cos t, \qquad x_2 = 0,$$

so that x_3 satisfies the equation

$$\ddot{x}_3 + x_3 = \tfrac{1}{6}\cos^3 t, \qquad x_3(0) = \dot{x}_3(0) = 0. \tag{11}$$

Solving (11) by the method of complex amplitudes, say, we get

$$x_3 = \alpha(\cos t - \cos 3t) + \beta t \sin t,$$

where $\alpha = 1/192$, $\beta = 1/16$.

Thus the effect of the nonlinearity ($\sin x \neq x$) on the oscillations of the pendulum reduces † to the presence of an extra term $A^3x_3 + O(A^4)$:

$$x = A \cos t + A^3[\alpha(\cos t - \cos 3t) + \beta t \sin t] + O(A^4).$$

The period T of the oscillations is just the point at which $x(t)$ has its maximum, and is near 2π for small A. To find this point, we use the condition $\dot{x}(T) = 0$:

$$A\{-\sin T + A^2[(\beta - \alpha)\sin T + 3\alpha \sin 3T + \beta T \cos T] + O(A^3)\} = 0. \tag{12}$$

To solve (12) approximately for small A, let $T = 2\pi + u$. This gives the equation

$$\sin u = A^2[2\pi\beta + O(u)] + O(A^3)$$

for u. By the implicit function theorem,

$$u = 2\pi\beta A^2 + O(A^3),$$

i.e.,

$$T = 2\pi\left[1 + \frac{A^2}{16} + o(A^2)\right],$$

where $o(A^2) = O(A^4)$ since $T(A)$ is even.

† Here it is useful to recall the bucket with the hole in its bottom (see the warning in Sec. 9.5). From the presence of the "secular" term $t \sin t$ in the formula for x_3, we can draw no conclusions whatsoever about the behavior of the pendulum as $t \to \infty$. Our approximation is valid only for a finite time interval, and the term $O(A^4)$ becomes large for large t. Thus our solution of the equation for oscillations of a pendulum actually remains bounded (by A) for all t, as is apparent from the law of conservation of energy.

Problem 2. Investigate the dependence of the period T of the oscillations on the amplitude A for the equation

$$\ddot{x} + \omega^2 x + ax^2 + bx^3 = 0.$$

Ans. $T = \dfrac{2\pi}{\omega}\left[1 + \left(\dfrac{5a^2}{12\omega^4} - \dfrac{3b}{8\omega^2}\right)A^2 + o(A^2)\right].$

Problem 3. Deduce the same result from the explicit formula for the period (Sec. 12.7).

27. Nonautonomous Linear Equations

That part of the theory of linear equations which does not depend on shift invariance can easily be carried over to linear equations and systems with variable coefficients.

27.1. Definition.   By a (*homogeneous*) *linear equation with variable coeffi-cients*† is meant an equation of the form

$$\dot{x} = A(t)x, \qquad x \in \mathbf{R}^n, \quad A(t) : \mathbf{R}^n \to \mathbf{R}^n, \tag{1}$$

where t belongs to an open interval I of the real axis (possibly the whole real axis).

Geometrically the solutions of equation (1) are represented by integral curves in the strip $I \times \mathbf{R}^n$ of extended phase space (Fig. 182). As usual, we will assume that the function $A(t)$ is smooth.‡

Example 1.   Consider the pendulum equation $\ddot{x} = -\omega^2 x$. The frequency ω is determined by the length of the pendulum, and the oscillations of a

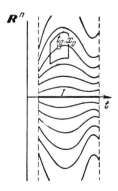

Fig. 182   Integral curves of a linear equation.

† Here we assume that the coefficients are real. The complex case is completely analogous.
‡ It is actually enough to assume that $A(t)$ is continuous (see Sec. 32.6).

pendulum of variable length are described by the analogous equation

$$\ddot{x} = -\omega^2(t)x.$$

This equation can be written in the form (1):

$$\begin{cases} \dot{x}_1 = x_2, \\ \dot{x}_2 = -\omega^2(t)x_1, \end{cases} \qquad A = \begin{pmatrix} 0 & 1 \\ -\omega^2(t) & 0 \end{pmatrix}.$$

The swing (Fig. 183) is an example of a pendulum of variable length. In fact, by varying the position of her center of gravity, the girl on the swing can periodically vary the value of the parameter ω.

27.2. Existence of solutions.   One solution of equation (1) is obviously the null solution. For an arbitrary initial condition $(t_0, \mathbf{x}_0) \in I \times \mathbf{R}^n$, the existence of a solution defined in a neighborhood of the point t_0 follows from the general theorems of Chap. 2. For a nonlinear equation it may not be possible to extend this solution onto the whole interval I (Fig. 184). However, linear equations have the special feature that none of their solutions can become infinite in a finite time interval.

THEOREM.   *Every solution of equation* (1) *can be extended onto the whole interval I.*

The idea of the proof is that $|\dot{\mathbf{x}}| \leqslant C|\mathbf{x}|$ for a linear equation, and hence the solution can grow no faster than e^{Ct}.

To give a rigorous proof, we proceed as follows, say, noting first that if $[a, b]$ is a compact interval in I, then the norm† of the operator $A(t)$ is bounded on $[a, b]$:

$$|A(t)| < C = C(a, b). \tag{2}$$

Fig. 183   The swing.

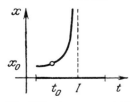

Fig. 184   A nonextendable solution of the equation $\dot{x} = x^2$.

† We assume that some Euclidean metric has been chosen in $\mathbf{R}^n$.

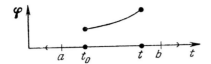

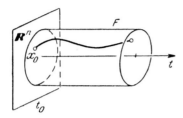

Fig. 185   A priori estimate of the growth of the solution on the interval $[a, b]$.

Fig. 186   Extension of a solution up to $t = b$.

LEMMA. *Let $\varphi(t)$ be a solution of equation* (1) *defined on the interval $[t_0, t]$, where $a \leqslant t_0 \leqslant t \leqslant b$* (Fig. 185). *Then $\varphi(t)$ satisfies the a priori estimate*

$$|\varphi(t)| \leqslant e^{C(t - t_0)}|\varphi(t_0)|. \tag{3}$$

Proof. The estimate is obvious for the null solution. If $\varphi(t_0) \neq 0$, then $\varphi(\tau) \neq 0$ by the uniqueness theorem. Let $r(\tau) = |\varphi(\tau)|$. Then the function $L(\tau) = \ln r^2$ is defined for $t_0 \leqslant \tau \leqslant t$. But

$$\dot{L} = \frac{2r\dot{r}}{r^2} \leqslant 2C$$

because of (2), and hence

$$L(t) \leqslant L(t_0) + 2C(t - t_0),$$

which implies (3). ∎

Proof of the theorem. Let $|\mathbf{x}_0|^2 = B > 0$, and consider the compact set

$$F = \{t, \mathbf{x}: a \leqslant t \leqslant b, |\mathbf{x}|^2 \leqslant 2Be^{2C(b-a)}\}$$

in extended phase space (Fig. 186). By the extension theorem, the solution with initial condition $\varphi(t_0) = \mathbf{x}_0$ can be extended forward up to the boundary of the cylinder F. The boundary of F consists of two end faces ($t = a, t = b$) and a lateral surface ($|\mathbf{x}|^2 = 2Be^{2C(b-a)}$). The solution cannot leave F on the lateral surface, since

$$|\varphi(t)|^2 \leqslant Be^{2C(b-a)}$$

by the lemma. Hence the solution can be extended to the right up to $t = b$. Similarly, it can be shown that the solution can be extended to the left up to $t = a$. Since a and b are arbitrary, the proof is now complete. ∎

27.3. The space of solutions of equation (1).

Let X be the set of all solutions of equation (1), defined on the whole interval I. Since solutions are just mappings $\varphi: I \to \mathbf{R}^n$ with values in the linear phase space $\mathbf{R}^n$, they can be added and multiplied by numbers:

$$(c_1\boldsymbol{\varphi}_1 + c_2\boldsymbol{\varphi}_2)(t) = c_1\boldsymbol{\varphi}_1(t) + c_2\boldsymbol{\varphi}_2(t).$$

THEOREM 1.   *The set X of all solutions of equation* (1) *defined on an interval I is a linear space.*

Proof.   Obvious, since

$$\frac{d}{dt}(c_1\boldsymbol{\varphi}_1 + c_2\boldsymbol{\varphi}_2) = c_1\dot{\boldsymbol{\varphi}}_1 + c_2\dot{\boldsymbol{\varphi}}_2 = c_1A\boldsymbol{\varphi}_1 + c_2A\boldsymbol{\varphi}_2 = A(c_1\boldsymbol{\varphi}_1 + c_2\boldsymbol{\varphi}_2). \quad \blacksquare$$

THEOREM 2.   *The linear space X of solutions of a linear equation is isomorphic to the phase space $\mathbf{R}^n$ of the equation.*

Proof.   Let $t \in I$, and consider the mapping

$$B_t: X \to \mathbf{R}^n, \qquad B_t\boldsymbol{\varphi} = \boldsymbol{\varphi}(t)$$

associating with every solution its value at time t. The mapping B_t is linear, since the value of a sum of solutions equals the sum of their values. The image of B_t is the whole phase space $\mathbf{R}^n$, since by the existence theorem, for every $\mathbf{x}_0 \in \mathbf{R}^n$ there exists a solution $\boldsymbol{\varphi}$ with initial condition $\boldsymbol{\varphi}(t_0) = \mathbf{x}_0$. Finally, the kernel of B_t equals $\{0\}$, since the solution with initial condition $\boldsymbol{\varphi}(t_0) = 0$ is identically zero, by the uniqueness theorem. $\blacksquare$

Thus *the mapping B_t is an isomorphism of X onto $\mathbf{R}^n$.* This is the basic result of the theory of linear equations.

Definition.   By a *fundamental system of solutions* of equation (1) is meant any basis of the linear solution space X.

Problem 1. Find a fundamental system of solutions of equation (1) with

$$A = \begin{pmatrix} 0 & 1 \\ -1 & 0 \end{pmatrix}.$$

Theorem 2 has a number of immediate consequences:

COROLLARY 1.   *Every equation* (1) *has a fundamental system of n solutions $\boldsymbol{\varphi}_1, \ldots, \boldsymbol{\varphi}_n$.*

COROLLARY 2.   *Every solution of equation* (1) *is a linear combination of solutions of a fundamental system.*

COROLLARY 3.   *Any $n + 1$ solutions of equation* (1) *are linearly dependent.*

COROLLARY 4.   *The (t_0, t_1)-advance mapping*

$$g_{t_0}^{t_1} = B_{t_1}B_{t_0}^{-1}: \mathbf{R}^n \to \mathbf{R}^n$$

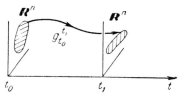

Fig. 187   The linear transformation of phase space produced by advancing the solutions of a linear equation from t_0 to t_1.

is a linear isomorphism (Fig. 187).

27.4. The Wronskian.   Let $\mathbf{e}_1, \ldots, \mathbf{e}_n$ be a basis in the phase space $\mathbf{R}^n$. The choice of a basis fixes a unit of volume and an orientation in $\mathbf{R}^n$, thereby assigning a definite volume to every parallelepiped in phase space.

Definition.   By the *Wronskian* (*determinant*) of a system of vector functions

$$\boldsymbol{\varphi}_k(t) : I \to \mathbf{R}^n, \qquad k = 1, \ldots, n$$

is meant the numerical function $W : I \to \mathbf{R}$ whose value at the point t equals the (oriented) volume of the parallelepiped spanned by the vectors $\boldsymbol{\varphi}_1(t), \ldots, \boldsymbol{\varphi}_n(t) \in \mathbf{R}^n$. Thus

$$W(t) = \begin{vmatrix} \varphi_{11}(t) & \cdots & \varphi_{n1}(t) \\ \cdot & \cdots & \cdot \\ \varphi_{1n}(t) & \cdots & \varphi_{nn}(t) \end{vmatrix},$$

where

$$\boldsymbol{\varphi}_k(t) = \varphi_{k1}(t)\mathbf{e}_1 + \cdots + \varphi_{kn}(t)\mathbf{e}_n.$$

In particular, let the $\boldsymbol{\varphi}_k$ be solutions of equation (1). Their images under the isomorphism B_t constructed above are vectors $\boldsymbol{\varphi}_k(t) \in \mathbf{R}^n$ in phase space. These vectors are linearly dependent if and only if the Wronskian vanishes at the point t. This implies

COROLLARY 5.   *A system of solutions* $\boldsymbol{\varphi}_1, \ldots, \boldsymbol{\varphi}_n$ *of equation* (1) *is fundamental if and only if its Wronskian is nonzero at some point.*

COROLLARY 6.   *If the Wronskian of a system of solutions of equation* (1) *vanishes at even one point, then it vanishes identically for all t.*

Problem 1. Can the Wronskian of a system of linearly independent vector functions $\boldsymbol{\varphi}_k$ vanish identically?

Problem 2. Prove that the Wronskian of a fundamental system of solutions is proportional to the determinant of the (t_0, t)-advance mapping:

$$W(t) = \det (g_{t_0}^t) \, W(t_0).$$

Hint. See Sec. 27.6.

27.5. The case of a single equation of order n.   Consider the homogeneous linear equation of order n

$$x^{(n)} + a_1 x^{(n-1)} + \cdots + a_n x = 0, \tag{4}$$

with coefficients $a_k = a_k(t)$, $t \in I$ which are in general variable.

Some second-order differential equations with variable coefficients are encountered so often in the applications that they have special names, and their solutions have been studied and tabulated in no less detail than the sine and cosine functions.†

Example 1. Bessel's equation

$$\ddot{x} + \frac{1}{t}\dot{x} + \left(1 - \frac{\nu^2}{t^2}\right)x = 0.$$

Example 2. Gauss' hypergeometric equation

$$\ddot{x} + \frac{(\alpha + \beta + 1)t - \gamma}{t(t-1)}\dot{x} + \frac{\alpha\beta}{t(t-1)}x = 0.$$

Example 3. Mathieu's equation

$$\ddot{x} + (a + b\cos t)\dot{x} = 0.$$

We could write equation (4) as a system of n first-order equations and then apply the preceding considerations. However, we prefer to consider the space X of solutions of equation (4) directly. The space X is a linear space of functions $\varphi: I \to \mathbf{R}$ which is naturally isomorphic to the space of solutions of the equivalent system of n equations. To specify the isomorphism, we assign each φ the vector functions

$$\boldsymbol{\varphi} = (\varphi, \dot{\varphi}, \ldots, \varphi^{(n-1)})$$

made up of the derivatives of φ:

COROLLARY 7.   *The space X of solutions of equation* (4) *is isomorphic to the phase space* $\mathbf{R}^n$ *of equation* (4), *where the isomorphism can be specified by assigning each* $\varphi \in X$ *the vector*

$$(\varphi(t_0), \dot{\varphi}(t_0), \ldots, \varphi^{(n-1)}(t_0))$$

made up of the derivatives of φ at some point t_0.

Definition.   By a *fundamental system of solutions* of equation (4) is meant any basis of the solution space X.

---

† See e.g., E. Jahnke and F. Emde, *Tables of Higher Functions*, sixth edition, revised by F. Lösch, McGraw-Hill, New York (1960).

Problem 1. Find a fundamental system of solutions of equation (4) for the case where the coefficients a_k are constant, e.g., for $\ddot{x} + ax = 0$.

Ans. $\{t^r e^{\lambda t}\}$, $0 \leqslant r < \nu$, where λ is a root of multiplicity ν of the characteristic equation. In the case of complex roots $\lambda = \alpha \pm i\omega$, we must change $e^{\lambda t}$ to $e^{\alpha t} \cos \omega t$, $e^{\alpha t} \sin \omega t$. In particular, for $\ddot{x} + ax = 0$ we have

$$\begin{cases} \cos \omega t, \sin \omega t & \text{if} \quad a = \omega^2 > 0, \\ \cosh \alpha t, \sinh \alpha t \\ \text{or } e^{\alpha t}, e^{-\alpha t} & \text{if} \quad a = -\alpha^2 < 0, \\ 1, t & \text{if} \quad a = 0. \end{cases}$$

Definition. By the *Wronskian* of a system of numerical functions

$$\varphi_k(t): I \to \mathbf{R}, \qquad k = 1, \ldots, n$$

is meant the numerical function $W: I \to \mathbf{R}$ whose value at the point t equals

$$W(t) = \begin{vmatrix} \varphi_1(t) & \cdots & \varphi_n(t) \\ \dot{\varphi}_1(t) & \cdots & \dot{\varphi}_n(t) \\ \cdot & \cdots & \cdot \\ \varphi_1^{(n-1)}(t) & \cdots & \varphi_n^{(n-1)}(t) \end{vmatrix}.$$

In other words, W is just the Wronskian of the system of vector functions $\boldsymbol{\varphi}_k(t): I \to \mathbf{R}^n$ obtained from the φ_k in the usual way:

$$\boldsymbol{\varphi}_k(t) = (\varphi_k(t), \dot{\varphi}_k(t), \ldots, \varphi_k^{(n-1)}(t)), \qquad k = 1, \ldots, n.$$

Everything said about the Wronskian of a system of vector solutions of equation (1) carries over without change to the Wronskian of a system of solutions of equation (4). In particular, we have

COROLLARY 8.   *If the Wronskian of a system of solutions of equation (4) vanishes at even one point, then it vanishes identically.*

Problem 2. Suppose the Wronskian of two functions vanishes at the point t_0. Does it follow that the Wronskian vanishes identically?

COROLLARY 9.   *If the Wronskian of a system of solutions of equation (4) vanishes at even one point, then the solutions are linearly dependent.*

Problem 3. Suppose the Wronskian of two functions vanishes identically. Does it follow that the functions are linearly dependent?

COROLLARY 10.   *A system of solutions $\varphi_1, \ldots, \varphi_n$ of equation (4) is fundamental if and only if its Wronskian is nonzero at some point.*

Example 4.   Consider the system of functions $e^{\lambda_1 t}, \ldots, e^{\lambda_n t}$. These functions form a fundamental system of solutions of a linear equation of the form (4) (which one?). Therefore they are linearly independent, so that their Wronskian is nonzero. But this determinant equals

$$
W = \begin{vmatrix} e^{\lambda_1 t} & \cdots & e^{\lambda_n t} \\ \lambda_1 e^{\lambda_1 t} & \cdots & \lambda_n e^{\lambda_n t} \\ & \cdots & \\ \lambda_1^{n-1} e^{\lambda_1 t} & \cdots & \lambda_n^{n-1} e^{\lambda_n t} \end{vmatrix} = e^{(\lambda_1 + \cdots + \lambda_n)t} \begin{vmatrix} 1 & \cdots & 1 \\ \lambda_1 & \cdots & \lambda_n \\ & \cdots & \\ \lambda_1^{n-1} & \cdots & \lambda_n^{n-1} \end{vmatrix}.
$$

COROLLARY 11.   *The Vandermonde determinant*

$$
\begin{vmatrix} 1 & \cdots & 1 \\ \lambda_1 & \cdots & \lambda_n \\ & \cdots & \\ \lambda_1^{n-1} & \cdots & \lambda_n^{n-1} \end{vmatrix}
$$

is nonzero if the numbers λ_k are distinct.

Example 5.   The pendulum equation $\ddot{x} + \omega^2 x = 0$ has $\cos \omega t$, $\sin \omega t$ as a fundamental system of solutions. The Wronskian

$$
W = \begin{vmatrix} \cos \omega t & \sin \omega t \\ -\omega \sin \omega t & \omega \cos \omega t \end{vmatrix} = \omega
$$

is constant. This is hardly surprising, since the phase flow of the pendulum equation preserves area (see Sec. 16.4).

27.6. Liouville's theorem.   We now examine how the volume of figures in phase space changes in the general case under the action of the transformation $g_{t_0}^t$ during the time from t_0 to t.

THEOREM **(Liouville).**   *The Wronskian of a system of solutions of equation* (1) *satisfies the differential equation*

$$
\dot{W} = aW, \qquad a(t) = \operatorname{Tr} A(t), \tag{5}
$$

involving the trace of the operator $A(t)$.

It follows from the theorem, which we will prove in a moment, that

$$
W(t) = \exp\left\{ \int_{t_0}^{t} a(\tau)\, d\tau \right\} W(t_0), \qquad \det g_{t_0}^t = \exp\left\{ \int_{t_0}^{t} a(\tau)\, d\tau \right\}. \tag{6}
$$

In fact, we can easily solve equation (5), obtaining

$$
\frac{dW}{W} = a\, dt, \qquad \ln W - \ln W_0 = \int_{t_0}^{t} a(\tau)\, d\tau.
$$

Incidentally, formula (6) again shows that the Wronskian of a system of solutions either vanishes identically or else does not vanish at all.

Problem 1. Find the volume of the image of the unit cube $0 \leqslant x_i \leqslant 1, i = 1, 2, 3$, under the

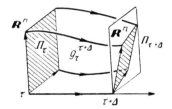

Fig. 188   Action of the phase flow on the parallelepiped Π_τ spanned by a fundamental system of solutions.

action of the transformation during time t of the phase flow of the system

$$\dot{x}_1 = 2x_1 - x_2 - x_3,$$
$$\dot{x}_2 = x_1 + x_2 + x_3,$$
$$\dot{x}_3 = x_1 - x_2 - x_3.$$

Ans. $W(t) = e^{2t}W(0) = e^{2t}$, since Tr $A = 2$.

The idea of the proof of Liouville's theorem is the following. If the coefficients are constant, the theorem reduces to Liouville's formula proved in Sec. 16.4. "Freezing" the coefficients $A(t)$, i.e., equating them to their values at some fixed instant of time τ, we can convince ourselves of the validity of equation (5) for arbitrary τ.

Proof of Liouville's theorem. Let

$$g_\tau^{\tau+\Delta} : \mathbf{R}^n \to \mathbf{R}^n$$

be the $(\tau, \tau + \Delta)$-advance mapping (Fig. 188), where Δ is small. This linear transformation of phase space carries the value of any solution $\boldsymbol{\varphi}$ of equation (1) at the time τ into its value at the time $\tau + \Delta$. According to (1),

$$\boldsymbol{\varphi}(\tau + \Delta) = \boldsymbol{\varphi}(\tau) + A(\tau)\boldsymbol{\varphi}(\tau)\Delta + o(\Delta),$$

i.e.,

$$g_\tau^{\tau+\Delta} = E + \Delta A(\tau) + o(\Delta).$$

Therefore, according to Sec. 16.1, the coefficient of volume expansion under the transformation $g_\tau^{\tau+\Delta}$ equals

$$\det g_\tau^{\tau+\Delta} = 1 + \Delta a + o(\Delta),$$

where $a = \text{Tr } A$. But $W(\tau)$ is the volume of the parallelepiped Π_τ spanned by the values of our system of solutions at the time τ, and the transformation $g_\tau^{\tau+\Delta}$ carries these values into the values of the same system of solutions at the time $\tau+\Delta$. The parallelepiped $\Pi_{\tau+\Delta}$ spanned by the new values has volume $W(\tau+\Delta)$. Therefore

$$W(\tau+\Delta) = \det (g_\tau^{\tau+\Delta})W(\tau) = [1 + a(\tau)\Delta + o(\Delta)]W(\tau),$$

which implies (5). ∎

It follows from Liouville's theorem that the Wronskian of the system of solutions of equation (4) equals

$$W(t) = \exp\left\{ -\int_{t_0}^{t} a_1(\tau)\, d\tau \right\} W(t_0).$$

Here the appearance of the minus sign stems from the fact that in writing
(4) in the form of a system (1), we must transpose $a_1 x^{(n-1)}$ to the right-hand
side. The matrix of the resulting system is

$$\begin{pmatrix} 0 & 1 & & \\ & \cdot & \cdot & \\ & & \cdot & 1 \\ -a_n & \cdots & & -a_1 \end{pmatrix},$$

with $-a_1$ as the only nonzero element on its main diagonal.

Example 1.   In the case of the swing, with equation

$$\ddot{x} + f(t)x = 0, \tag{7}$$

the equilibrium position $x = \dot{x} = 0$ cannot be asymptotically stable for any choice of
$f(t)$. In fact, consider any basis ξ, η in the plane $\mathbf{R}^2$ of the initial conditions
(Fig. 189). Stability means that $g_{t_0}^t \xi \to 0$, $g_{t_0}^t \eta \to 0$, in which case $W(t) \to 0$
for the corresponding fundamental system. But (7) is equivalent to the
system

$$\begin{cases} \dot{x}_1 = x_2, \\ \dot{x}_2 = -f(t)x_1, \end{cases}$$

with matrix

$$A = \begin{pmatrix} 0 & 1 \\ -f & 0 \end{pmatrix}.$$

Since $\operatorname{Tr} A = 0$, it follows that $W(t) = \text{const}$, contrary to $W \to 0$.

Problem 2. Consider the swing with friction

$$\ddot{x} + \alpha(t)\dot{x} + \omega^2(t)x = 0.$$

Show that asymptotic stability is impossible if the coefficient of friction is negative, i.e.,

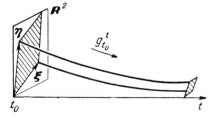

Fig. 189   The phase flow of an asymptotically stable linear system.

if $\alpha(t) < 0$ for all t. Is it true that the equilibrium position $(0, 0)$ is always stable if the coefficient of friction is positive?

Definition. By the *divergence* of a vector field $\mathbf{v}$ in the Euclidean space $\mathbf{R}^n$ with rectangular coordinates x_i is meant the function

$$\text{div } \mathbf{v} = \sum_{i=1}^{n} \frac{\partial v_i}{\partial x_i}.$$

In particular, *for a linear vector field* $\mathbf{v}(x) = A\mathbf{x}$, *the divergence is just the trace of the operator* A:

$$\text{div } A\mathbf{x} = \text{Tr } A.$$

The divergence of a vector field determines the rate of volume expansion due to the corresponding phase flow.

Let D be a domain in the Euclidean phase space of the (not necessarily linear) equation $\dot{\mathbf{x}} = \mathbf{v}(\mathbf{x})$, let $D(t)$ denote the image of D under the action of the phase flow, and let $V(t)$ denote the volume of the domain $D(t)$.

*Problem 3.   Prove the following stronger version of *Liouville's theorem*:

$$\frac{dV}{dt} = \int_{D(t)} \text{div } \mathbf{v} \, dx$$

(Fig. 190).

COROLLARY 1.   *If* $\text{div } \mathbf{v} = 0$, *then the phase flow preserves the volume of any domain.*

Such a phase flow can be thought of as the flow of an incompressible "phase fluid" in phase space.

COROLLARY 2.   *The phase flow of Hamilton's equations*

$$\dot{p}_k = -\frac{\partial H}{\partial q_k}, \qquad \dot{q}_k = \frac{\partial H}{\partial p_k}, \qquad k = 1, \ldots, n$$

preserves volume.

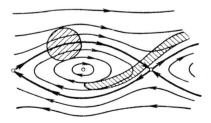

Fig. 190   The phase flow of a vector field of divergence zero preserves area.

Proof. Merely note that

$$\text{div } \mathbf{v} = \sum_{k=1}^{n} \left(\frac{\partial^2 H}{\partial q_k \partial p_k} - \frac{\partial^2 H}{\partial p_k \partial q_k} \right) = 0. \quad \blacksquare$$

This fact plays a key role in statistical physics.

28. Linear Equations with Periodic Coefficients

The theory of linear equations with periodic coefficients shows how to "pump up" a swing and explains why the upper equilibrium position of a pendulum, which is usually unstable, becomes stable if the point of suspension of the pendulum executes sufficiently rapid oscillations in the vertical direction.

28.1. The period-advance mapping.   Consider the differential equation

$$\dot{\mathbf{x}} = \mathbf{v}(\mathbf{x}, t), \qquad \mathbf{v}(\mathbf{x}, t + T) = \mathbf{v}(\mathbf{x}, t), \qquad \mathbf{x} \in \mathbf{R}^n, \tag{1}$$

whose right-hand side depends periodically on time (Fig. 191).

Example 1.   The motion of a pendulum with periodically varying parameters (for example, the motion of a swing) is described by a system of equations of the form (1):

$$\begin{cases} \dot{x}_1 = x_2, \\ \dot{x}_2 = -\omega^2(t)x_1, \end{cases} \qquad \omega(t + T) = \omega(t). \tag{2}$$

We will assume that all the solutions of equation (1) can be extended indefinitely. This is certainly true for the linear equations in which we are particularly interested.

The periodicity of the right-hand side of (1) leads to a number of special properties of the corresponding phase flow.

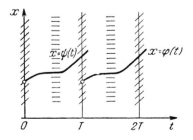

Fig. 191   The extended phase space of an equation with periodic coefficients.

LEMMA 1.   *The (t_1, t_2)-advance mapping $g_{t_1}^{t_2}: \mathbf{R}^n \to \mathbf{R}^n$ of phase space does not change when both t_1 and t_2 are increased by the period T of the right-hand side of* (1).

Proof.   We must show that the shift $\boldsymbol{\psi}(t) = \boldsymbol{\varphi}(t + T)$ of a solution of $\boldsymbol{\varphi}(t)$ by the time T is itself a solution. But a shift of the extended phase space by T along the time axis carries the direction field of equation (1) into itself (Fig. 191). Therefore an integral curve of (1) shifted by T is still everywhere tangential to the direction field, and hence remains an integral curve. It follows that

$$g_{t_1 + T}^{t_2 + T} = g_{t_1}^{t_2}. \quad \blacksquare$$

In particular, consider the transformation g_0^T produced by the phase flow during one period T. This "period-advance" mapping, which we denote by

$$A = g_0^T: \mathbf{R}^n \to \mathbf{R}^n$$

(Fig. 192), will play an important role in the considerations that follow.

Example 2.   For the systems

$$\begin{cases} \dot{x}_1 = x_2, \\ \dot{x}_2 = -x_1, \end{cases} \qquad \begin{cases} \dot{x}_1 = x_1, \\ \dot{x}_2 = -x_2, \end{cases}$$

which can be regarded as periodic with any period T, the mapping A is a rotation and a hyperbolic rotation respectively.

LEMMA 2.   *The transformations g_0^{nT} form a group*

$$g_0^{nT} = A^n,$$

and moreover

$$g_0^{nT + s} = g_0^s g_0^{nT}.$$

Proof.   By Lemma 1,

$$g_{nT}^{nT + s} = g_0^s,$$

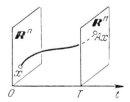

Fig. 192   The period-advance mapping.

and hence

$$g_0^{nT+s} = g_{nT}^{nT+s} g^{nT} = g_0^s g_0^{nT}.$$

Setting $s = T$, we get

$$g_0^{(n+1)T} = A g_0^{nT},$$

and hence $g_0^{nT} = A^n$ by induction. ∎

To every property of the solutions of equation (1) there corresponds an analogous property of the period-advance mapping A.

THEOREM.

1)   *A point $\mathbf{x}_0$ is a fixed point of the mapping $A (A\mathbf{x}_0 = \mathbf{x}_0)$ if and only if the solution with initial condition $\mathbf{x}(0) = \mathbf{x}_0$ is periodic with period T.*

2)   *A periodic solution $\mathbf{x}(t)$ is stable in Lyapunov's sense (asymptotically stable) if and only if the fixed point $\mathbf{x}_0$ of the mapping A is stable in Lyapunov's sense (asymptotically stable).*†

3)   *If the system (1) is linear, i.e., if $\mathbf{v}(\mathbf{x}, t) = \mathbf{v}(t)\mathbf{x}$ is a linear function of $\mathbf{x}$, then the mapping A is linear.*

4)   *If, moreover, the trace of the linear operator $V(t)$ vanishes, then the mapping A conserves volume:* $\det A = 1$.

Proof.   Assertions 1) and 2) follow from the condition $g_0^{T+s} = g_0^s A$ and from the continuous dependence of the solution on the initial conditions in the interval $[0, T]$. Assertion 3) follows from the fact that a sum of solutions of a linear system is itself a solution, while assertion 4) follows from Liouville's theorem. ∎

28.2. Stability conditions.   We now apply the above theorem to the mapping A of the phase plane (x_1, x_2) onto itself corresponding to the system (2). Since the system (2) is linear and the trace of the matrix of its right-hand side vanishes, we have the following

COROLLARY.   *The mapping A is linear and preserves area $(\det A = 1)$. The null solution of the system of equations (2) is stable if and only if the mapping A is stable.*

Problem 1. Prove that a rotation of the plane is a stable mapping, while a hyperbolic rotation is unstable.

We now make a more detailed study of linear mappings of the plane onto itself which preserve area.

---

† A fixed point $\mathbf{x}_0$ of the mapping A is said to be *stable in Lyapunov's sense* if $\forall\ \varepsilon > 0\ \exists\ \delta > 0$ such that $|\mathbf{x} - \mathbf{x}_0| < \delta$ implies $|A^n\mathbf{x} - A^n\mathbf{x}_0| < \varepsilon$ for all $n = 1, 2, \ldots$ and *asymptotically stable* if $A^n\mathbf{x} - A^n\mathbf{x}_0 \to 0$ as $n \to \infty$.

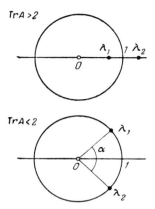

Fig. 193   Eigenvalues of the period-advance mapping.

THEOREM.   *Let A be a linear area-preserving mapping of the plane onto itself* (det $A = 1$). *Then the mapping A is stable if $|\mathrm{Tr}\,A| < 2$ and unstable if $|\mathrm{Tr}\,A| > 2$.*

Proof.   Let λ_1 and λ_2 be the eigenvalues of A, satisfying the characteristic equation

$$\lambda^2 - \lambda\,\mathrm{Tr}\,A + 1 = 0$$

with real coefficients

$$\lambda_1 + \lambda_2 = \mathrm{Tr}\,A, \qquad \lambda_1\lambda_2 = \det A = 1.$$

The roots λ_1 and λ_2 of the characteristic equation are real if $|\mathrm{Tr}\,A| > 2$ and complex conjugates if $|\mathrm{Tr}\,A| < 2$ (Fig. 193). In the first case, one of the eigenvalues has absolute value greater than 1 and the other absolute value less than 1, so that A is a hyperbolic rotation and hence unstable. In the second case, the eigenvalues lie on the unit circle:

$$\lambda_1\lambda_2 = \lambda_1\bar{\lambda}_1 = |\lambda_1|^2 = 1.$$

Hence the mapping A is equivalent to a rotation through the angle α (where $\lambda_{1,2} = e^{\pm i\alpha}$), i.e., A is a rotation for a suitable choice of a Euclidean structure in the plane (why?) and hence stable.   ∎

Thus the whole question of the stability of the null solution of the system (2) reduces to calculating the trace of the matrix A. Unfortunately, the trace can be calculated explicitly only in special cases. The trace can always be found approximately by numerical integration of the equation in the interval $0 \leqslant t \leqslant T$. In the important case where $\omega(t)$ is almost constant, some simple general considerations are useful.

28.3. Strongly stable systems.   Consider a linear system (1) with a two-dimensional phase space (i.e., with $n = 2$). Then (1) is said to be a *Hamiltonian system* if the divergence of **v** vanishes. As noted above, the phase flow of a Hamiltonian system conserves area: $\det A = 1$.

Definition.   The null solution of a linear Hamiltonian system is said to be *strongly stable* if it is stable and if the null solution of every neighboring linear Hamiltonian system is also stable.

The preceding two theorems now imply the following

COROLLARY.   *The null solution is strongly stable if* $|\text{Tr } A| < 2$.

Proof.   If $|\text{Tr } A| < 2$, then $|\text{Tr } \tilde{A}| < 2$ for the mapping $\tilde{A}$ corresponding to any system "sufficiently near" the original system.   ∎

We now apply this result to a system with almost constant coefficients. Consider, for example, the equation

$$\ddot{x} = -\omega^2[1 + \varepsilon a(t)]x, \qquad \varepsilon \ll 1, \tag{3}$$

where $a(t + 2\pi) = a(t)$, say $a(t) = \cos t$ (a pendulum whose frequency oscillates about ω with small amplitude and period 2π).† Every system (3) can be represented by a point in the plane of the parameters ω and ε (Fig. 194). Obviously the stable systems with $|\text{Tr } A| < 2$ form an open set in the plane (ω, ε) and the same is true of the unstable systems with $|\text{Tr } A| > 2$, while the "boundary of instability" is the set with equation $|\text{Tr } A| = 2$.

THEOREM.   *Every point of the ω-axis, except the points*

$$\omega = \frac{k}{2}, \qquad k = 0, 1, 2, \ldots$$

with integral and half-integral coordinates, corresponds to a strongly stable system (3).

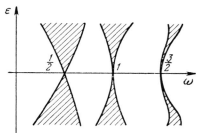

Fig. 194   The region of instability for parametric resonance.

† In the case $a(t) = \cos t$, equation (3) is called *Mathieu's equation*.

Thus the set of unstable systems can approach the ω-axis only at the points $\omega = k/2$. In other words, a swing can be "pumped up" by making a small periodic change in its length only in the case where the period of change of the length is near an integral number of half-periods of the natural frequency, a result everybody knows from experiment.

The proof of the theorem is based on the fact that for $\varepsilon = 0$, equation (3) has constant coefficients and can easily be solved.

Problem 1. Find the matrix of the period-advance mapping A for the system (3) with $\varepsilon = 0$ in the basis $x, \dot{x}$.

Solution. The general solution is

$$x = C_1 \cos \omega t + C_2 \sin \omega t,$$

so that

$$x = \cos \omega t, \qquad \dot{x} = -\omega \sin \omega t$$

is the particular solution satisfying the initial condition $x = 1, \dot{x} = 0$, while

$$x = \frac{1}{\omega} \sin \omega t, \qquad \dot{x} = \cos \omega t$$

is the particular solution satisfying the initial condition $x = 0, \dot{x} = 1$. *Ans.*

$$A = \begin{pmatrix} \cos 2\pi\omega & \frac{1}{\omega} \sin 2\pi\omega \\ -\omega \sin 2\pi\omega & \cos 2\pi\omega \end{pmatrix}.$$

Proof of the theorem. Note that $|\mathrm{Tr}\, A| = |2 \cos 2\pi\omega| < 2$ if $\omega \neq k/2$, $k = 0, 1, \ldots$ ∎

A more careful analysis[†] shows that quite generally (and, in particular, for $a(t) = \cos t$), the region of instability (the shaded region in Fig. 194) approaches the ω-axis near the points $\omega = k/2$, $k = 1, 2, \ldots$ Thus for certain ratios of the frequency of the change of parameters to the natural frequency of the swing ($\omega \approx k/2$, $k = 1, 2, \ldots$), the lower equilibrium position of the idealized swing is unstable, and it can be "pumped up" by an arbitrarily small periodic change of length. This phenomenon is known as "parametric resonance." The characteristic feature of parametric resonance is that it becomes strongest when the frequency ν of change of the parameters ($\nu = 1$ in equation (3)) is twice as large as the natural frequency ω.

Remark. In theory, parametric resonance is observed for infinitely many ratios $\omega/\nu \approx k/2$, $k = 1, 2, \ldots$, but the only cases usually observed in practice are those where k is small ($k = 1, 2$, less often 3). The point is that

[†] For example, see Problem 2 solved below.

a) For large k the region of instability approaches the ω-axis with a narrow tongue, and for the resonance frequency ω we have very narrow limits ($\sim \varepsilon^k$ for a smooth function $a(t)$ in (3)).

b) The instability itself is weak for large k, since the quantity $|\operatorname{Tr} A| - 2$ is small and the eigenvalues are near 1 for large k.

c) Even a slight amount of friction leads to the presence of a minimum value ε_k of the amplitude necessary for the occurrence of kth-order parametric resonance, with the oscillations being damped out for smaller values of ε_k. Moreover, ε_k grows rapidly with k (Fig. 195).

It should also be noted that x becomes arbitrarily large for equation (3) in the case of instability. In actual systems, the oscillations attains only a finite amplitude, since the linearized equation (3) itself becomes meaningless for large x and we must take account of nonlinear effects.

Problem 2. Find the form of the region of instability in the plane (ω, k) for the system described by the equation

$$\ddot{x} = -f(t)x, \tag{4}$$

where

$$f(t) = \begin{cases} \omega + \varepsilon, & 0 \leqslant t < \pi, \\ \omega - \varepsilon, & \pi \leqslant t < 2\pi, \end{cases} \quad \varepsilon \ll 1, \tag{4'}$$

$$f(t + 2\pi) = f(t).$$

Solution. It follows from the solution of Problem 1 that $A = A_2 A_1$, where

$$A_k = \begin{pmatrix} c_k & \dfrac{1}{\omega_k} s_k \\ -\omega_k s_k & c_k \end{pmatrix}, \qquad c_k = \cos \pi \omega_k, \qquad s_k = \sin \pi \omega_k, \qquad \omega_{1,2} = \omega \pm \varepsilon.$$

Hence the boundary of the zone of instability has the equation

$$|\operatorname{Tr} A| = \left| 2c_1 c_2 - \left(\frac{\omega_1}{\omega_2} + \frac{\omega_2}{\omega_1} \right) s_1 s_2 \right| = 2. \tag{5}$$

Since $\varepsilon \ll 1$, we have

$$\frac{\omega_2}{\omega_1} = \frac{\omega + \varepsilon}{\omega - \varepsilon} \approx 1.$$

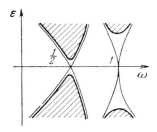

Fig. 195   Influence of slight friction on the region of instability.

Let Δ be such that

$$\frac{\omega_1}{\omega_2} + \frac{\omega_2}{\omega_1} = 2(1 + \Delta).$$

Then an easy calculation gives

$$\Delta = \frac{2\varepsilon^2}{\omega^2} + O(\varepsilon^4) \ll 1. \tag{6}$$

Using the formulas

$$2c_1 c_2 = \cos 2\pi\varepsilon + \cos 2\pi\omega,$$
$$2s_1 s_2 = \cos 2\pi\varepsilon - \cos 2\pi\omega,$$

we rewrite equation (5) in the form

$$-\Delta \cos 2\pi\varepsilon + (2 + \Delta)\cos 2\pi\omega = \pm 2$$

or

$$\cos 2\pi\omega = \frac{2 + \Delta \cos 2\pi\varepsilon}{2 + \Delta}, \tag{7}$$

$$\cos 2\pi\omega = \frac{-2 + \Delta \cos 2\pi\varepsilon}{2 + \Delta}. \tag{7'}$$

In the first case $\cos 2\pi\omega \approx 1$, and hence we write $\omega = k + a$, $|a| \ll 1$,

$$\cos 2\pi\omega = \cos 2\pi a = 1 - 2\pi^2 a^2 + O(a^4).$$

Thus, rewriting (7) as

$$\cos 2\pi\omega = 1 - \frac{\Delta}{2 + \Delta}(1 - \cos 2\pi\varepsilon),$$

we have

$$2\pi^2 a^2 + O(a^4) = \Delta\pi^2\varepsilon^2 + O(\varepsilon^4). \tag{8}$$

Substituting (6) into (8), we finally get

$$a = \pm\frac{\varepsilon^2}{\omega} + o(\varepsilon^2),$$

i.e.,

$$\omega = k \pm \frac{\varepsilon^2}{k} + o(\varepsilon^2)$$

(Fig. 196). Solving (7') in the same way, we get

$$\omega = k + \frac{1}{2} \pm \frac{\varepsilon}{\pi(k + \frac{1}{2})} + o(\varepsilon).$$

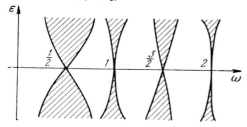

Fig. 196   The region of instability for equation (4).

Problem 3. Can the upper (usually unstable) equilibrium position of a pendulum become stable if the point of suspension oscillates in the vertical direction?

Ans. The upper equilibrium position becomes stable for sufficiently rapid oscillations of the point of suspension.

Solution. Let l be the length of the pendulum and $a \ll l$ the amplitude of the oscillations of the point of suspension. Let the period of the oscillations of the point of suspension be 2τ, where the acceleration of the point of suspension is constant and equal to $\pm c$ during every half-period (then $c = 8a/\tau^2$). The equation of motion can be written in the form

$$\ddot{x} = (\omega^2 \pm \alpha^2)x,$$

where the sign changes after the time τ and $\omega^2 = g/l$, $\alpha^2 = c/l$. If the oscillations of the point of suspension are sufficiently rapid, then $\alpha^2 > \omega^2$, where $\alpha^2 = 8a/l\tau^2$. As in the preceding problem, we have $A = A_2 A_1$, where

$$A_1 = \begin{pmatrix} \cosh k\tau & \dfrac{1}{k}\sinh k\tau \\ k\sinh k\tau & \cosh k\tau \end{pmatrix}, \qquad k^2 = \alpha^2 + \omega^2$$

and

$$A_2 = \begin{pmatrix} \cos \Omega\tau & \dfrac{1}{\Omega}\sin \Omega\tau \\ -\Omega\sin \Omega\tau & \cos \Omega\tau \end{pmatrix}, \qquad \Omega^2 = \alpha^2 - \omega^2.$$

Hence the stability condition $|\mathrm{Tr}\, A| < 2$ takes the form

$$\left| 2\cosh k\tau \cos \Omega\tau + \left(\frac{k}{\Omega} - \frac{\Omega}{k}\right)\sinh k\tau \sin \Omega\tau \right| < 2. \tag{9}$$

We now show that this condition holds for sufficiently rapid oscillations of the point of suspension, i.e., for $c \gg g$ ($\tau \ll 1$). Introducing dimensionless variables ε and μ such that

$$\frac{a}{l} = \varepsilon^2 \ll 1, \qquad \frac{g}{c} = \mu^2 \ll 1,$$

we have

$$k\tau = 2\sqrt{2}\varepsilon\sqrt{1 + \mu^2}, \qquad \tau = 2\sqrt{2}\varepsilon\sqrt{1 - \mu^2},$$
$$\frac{k}{\Omega} - \frac{\Omega}{k} = \sqrt{\frac{1 + \mu^2}{1 - \mu^2}} - \sqrt{\frac{1 - \mu^2}{1 + \mu^2}} = 2\mu^2 + O(\mu^4).$$

Therefore the expansions

$$\cosh k\tau = 1 + 4\varepsilon^2(1 + \mu^2) + \tfrac{8}{3}\varepsilon^4 + \cdots,$$
$$\cos \Omega\tau = 1 - 4\varepsilon^2(1 - \mu^2) + \tfrac{8}{3}\varepsilon^4 + \cdots,$$
$$\left(\frac{k}{\Omega} - \frac{\Omega}{k}\right)\sinh k\tau \sin \Omega\tau = 16\varepsilon^2\mu^2 + \cdots$$

are valid for small ε and μ with accuracy $O(\varepsilon^4 + \mu^4)$. Thus the stability condition (9) becomes

$$2\left(1 - 16\varepsilon^4 + \frac{16}{3}\varepsilon^4 + 8\varepsilon^2\mu^2 + \cdots\right) + 16\varepsilon^2\mu^2 < 2.$$

Neglecting small quantities of higher order, we find that

$$\mu < \sqrt{\frac{2}{3}} \varepsilon$$

or

$$\frac{g}{c} < \frac{2}{3}\frac{a}{l}.$$

This condition can be written in the form

$$N > \sqrt{\frac{3}{64}}\,\omega\frac{l}{a} \approx 0.2165\frac{l}{a},$$

where $N = 1/2\tau$ is the frequency of oscillation of the point of suspension. For example, if the length of the pendulum is $l = 20$ cm and the point of suspension executes oscillations of amplitude $a = 1$ cm, then $N > 0.2165\sqrt{980/20}\cdot 20$ cps. In particular, the upper equilibrium position is stable if the frequency of oscillation of the point of suspension exceeds 35, say.

29. Variation of Constants

The following method is often useful in investigating equations near "unperturbed" equations that have already been studied. Let c be a first integral of the unperturbed equation. Then c is no longer a first integral of the neighboring "perturbed" equations. However, it is often possible to recognize (exactly or approximately) how the values $c(\varphi(t))$ vary with time, where φ is the solution of the unperturbed equation. In particular, suppose the original equation is linear and homogeneous, while the perturbed equation is nonhomogeneous. Then this method leads to an explicit formula for the solution, where, because of the linearity, the perturbation need not satisfy any "smallness" requirement.

   We begin by noting that the particularly simple nonhomogeneous linear equation

$$\dot{\mathbf{x}} = f(t), \qquad \mathbf{x} \in \mathbf{R}^n, \qquad t \in I, \tag{1}$$

corresponding to the "simplest" homogeneous equation

$$\dot{\mathbf{x}} = 0, \tag{2}$$

can be solved by quadratures:

$$\boldsymbol{\varphi}(t) = \boldsymbol{\varphi}(t_0) + \int_{t_0}^{t} \mathbf{f}(\tau)\, d\tau. \tag{3}$$

29.1. The general case.   More generally, consider the nonhomogeneous linear equation

$$\dot{\mathbf{x}} = A(t)\mathbf{x} + \mathbf{h}(t), \qquad \mathbf{x} \in \mathbf{R}^n, \quad t \in I, \tag{4}$$

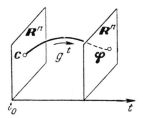

Fig. 197   The coordinates of the point **c** are first integrals of the homogeneous equation.

corresponding to the homogeneous equation

$$\dot{\mathbf{x}} = A(t)\mathbf{x}. \tag{5}$$

Suppose we know how to solve (5) and $\mathbf{x} = \boldsymbol{\varphi}(t)$ is its solution. Then in the extended phase space we use coordinates rectifying the integral curves of (5), i.e., the point $(\boldsymbol{\varphi}(t), t)$ is assigned the coordinates $\mathbf{c} = \boldsymbol{\varphi}(t_0)$ and t (Fig. 197). Equation (5) takes the particularly simple form (2) in the new coordinates, and we can go over to the rectifying coordinates by making a transformation linear in $\mathbf{x}$. Hence the nonhomogeneous equation (4) takes the particularly simple form (1) in the new coordinates, and can easily be solved.

29.2. Solution of equation (4).   Suppose we look for a solution of the nonhomogeneous equation (4) of the form

$$\boldsymbol{\varphi}(t) = g^t\mathbf{c}(t), \qquad \mathbf{c}: I \to \mathbf{R}^n, \tag{6}$$

where $g^t: \mathbf{R}^n \to \mathbf{R}^n$ is the linear (t_0, t)-advance mapping for the homogeneous equation (5). Differentiating (6) with respect to t, we get

$$\dot{\boldsymbol{\varphi}} = \dot{g}^t\mathbf{c} + g^t\dot{\mathbf{c}} = Ag^t\mathbf{c} + g^t\dot{\mathbf{c}} = A\boldsymbol{\varphi} + g^t\dot{\mathbf{c}},$$

which gives

$$g^t\dot{\mathbf{c}} = \mathbf{h}(t)$$

after substitution into (4). This proves the following

THEOREM.   *Formula* (6) *gives the solution of equation* (4) *if and only if* **c** *satisfies the equation*

$$\dot{\mathbf{c}} = \mathbf{f}(t), \tag{7}$$

where $\mathbf{f}(t) = (g^t)^{-1}\mathbf{h}(t)$.

COROLLARY.   *The solution of the nonlinear equation* (4) *with initial condition*

$\varphi(t_0) = \mathbf{c}$ *is given by*

$$\varphi(t) = g^t\left(\mathbf{c} + \int_{t_0}^t (g^\tau)^{-1}\mathbf{h}(\tau)\,d\tau\right).$$

Proof. Apply formula (3) to equation (7), which is of the particularly simple form (1). ∎

Remark. In coordinate form the theorem goes as follows: *Given a fundamental system of solutions of the homogeneous equation* (5), *the nonhomogeneous equation* (4) *can be solved by substituting a linear combination of solutions of the fundamental system into the nonhomogeneous equation and regarding the coefficients of the linear combination as unknown functions of time. The resulting equation for determining the coefficients is then of the particularly simple form* (1).

Problem 1. Solve the equation $\ddot{x} + x = f(t)$.

Solution. Form the corresponding homogeneous system

$$\begin{cases} \dot{x}_1 = x_2, \\ \dot{x}_2 = -x_1, \end{cases}$$

with the known system of fundamental solutions $x_1 = \cos t$, $x_2 = -\sin t$ and $x_1 = \sin t$, $x_2 = \cos t$. In accordance with the general rule, we look for a solution of the form

$$x_1 = c_1(t)\cos t + c_2(t)\sin t, \qquad x_2 = -c_1(t)\sin t + c_2(t)\cos t.$$

To determine c_1 and c_2, we have the system

$$\dot{c}_1 \cos t + \dot{c}_2 \sin t = 0, \qquad -\dot{c}_1 \sin t + \dot{c}_2 \cos t = f(t).$$

Therefore

$$\dot{c}_1 = -f(t)\sin t, \qquad \dot{c}_2 = f(t)\cos t,$$

so that finally

$$x(t) = \left[x(0) - \int_0^t f(\tau)\sin\tau\,d\tau\right]\cos t + \left[x(0) + \int_0^t f(\tau)\cos\tau\,d\tau\right]\sin t.$$

4 Proofs of the Basic Theorems

In this chapter we will prove the theorems on existence, uniqueness, continuity, and differentiability of ordinary differential equations, as well as the theorems on rectifiability of a vector field and of a field of directions. The proofs also contain a technique for constructing approximate solutions of differential equations.

30. Contraction Mappings

We now give a method for finding a fixed point of a mapping of a metric space into itself. This method will be used later to construct solutions of differential equations.

30.1. Definition. Let $A: M \to M$ be a mapping of a metric space M (with metric ρ) into itself. Then M is said to be a *contraction mapping* if there exists a constant λ, $0 < \lambda < 1$ such that

$$\rho(Ax, Ay) \leqslant \lambda \rho(x, y) \quad \forall \, x, y \in M. \tag{1}$$

Example 1. Let $A: \mathbf{R} \to \mathbf{R}$ be a real function of a real variable (Fig. 198). If the derivative of A is everywhere of absolute value less than 1, then A need not be a contraction mapping. However A is a contraction mapping if

$$|A'| \leqslant \lambda < 1.$$

Example 2. Let $A: \mathbf{R}^n \to \mathbf{R}^n$ be a linear operator. If all the eigenvalues of A lie strictly inside the unit disk, then there exists a Euclidean metric (a Lyapunov function in the sense of Sec. 22.3) such that A is a contraction mapping.

Problem 1. Which of the following mappings of the line (with the ordinary metric) into itself are contraction mappings:

a) $y = \sin x$;   b) $y = \sqrt{x^2 + 1}$;   c) $y = \arctan x$?

Problem 2. Can $\leqslant$ be replaced by $<$ in the inequality (1)? No, since $x=y$ always gives an equality.

30.2. The contraction mapping theorem. A point $x \in M$ is called a *fixed point* of the mapping $A: M \to M$ if $Ax = x$.

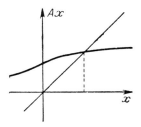

Fig. 198   Fixed point of a contraction mapping.

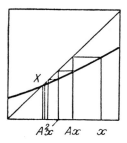

Fig. 199   Sequence of images of a point x under a mapping A.

THEOREM.   *Let* $A: M \to M$ *be a contraction mapping of a complete metric space* M *into itself. Then* A *has a unique fixed point. Given any point* $x \in M$, *the sequence*

$$x, Ax, A^2x, A^3x, \ldots$$

of images of x *under application of the operator* A (Fig. 199) *converges to the fixed point.*

Proof.   If $\rho(x, Ax) = d$, then

$$\rho(A^nx, A^{n+1}x) \leqslant \lambda^nd.$$

The series

$$\sum_{n=0}^{\infty} \lambda^n$$

converges, and hence the sequence A^nx, $n = 0, 1, 2, \ldots$ is a Cauchy sequence. But the space M is complete, and hence the limit

$$X = \lim_{n \to \infty} A^nx$$

exists. The point X is a fixed point of A. In fact, since every contraction mapping is continuous (choose $\delta = \varepsilon$), we have

$$AX = A \lim_{n \to \infty} A^nx = \lim_{n \to \infty} A^{n+1}x = X.$$

Moreover every fixed point Y coincides with X, since

$$\rho(X, Y) = \rho(AX, AY) \leqslant \lambda\rho(X, Y), \lambda < 1 \Rightarrow \rho(X, Y) = 0. \quad \blacksquare$$

Remark.   The points $x, Ax, A^2x, \ldots$ are called *successive approximations* to X.

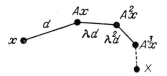

Fig. 200   Estimate of the accuracy of an approximation x to the fixed point X.

Let x be an approximation to the fixed point X of a contraction mapping A. Then the accuracy of the approximation is easily estimated in terms of the distance d between the points x and Ax. In fact

$$\rho(x, X) \leqslant \frac{d}{1 - \lambda},$$

since

$$d + \lambda d + \lambda^2 d + \cdots = \frac{d}{1 - \lambda}$$

(Fig. 200).

31. The Existence, Uniqueness, and Continuity Theorems

We now construct a contraction mapping of a complete metric space whose fixed point determines the solution of a given differential equation.

31.1. Successive Picard approximations.   Consider the differential equation $\dot{x} = \mathbf{v}(\mathbf{x}, t)$ determined by a vector field $\mathbf{v}$ in a domain of the extended phase space $\mathbf{R}^{n+1}$ (Fig. 201). Then by the *Picard mapping* we mean the following mapping of the function $\boldsymbol{\varphi}: t \mapsto \mathbf{x}$ into the function $A\boldsymbol{\varphi}: t \mapsto \mathbf{x}$ defined by

$$(A\boldsymbol{\varphi})(t) = \mathbf{x}_0 + \int_{t_0}^{t} \mathbf{v}(\boldsymbol{\varphi}(\tau), \tau)d\tau.$$

Geometrically the transition from $\boldsymbol{\varphi}$ to $A\boldsymbol{\varphi}$ (Fig. 202) means using one curve $\boldsymbol{\varphi}$ to construct a new curve $A\boldsymbol{\varphi}$ whose tangent at every point t is parallel to the direction field determined by $\boldsymbol{\varphi}$ rather than to the field on the new curve $A\boldsymbol{\varphi}$ itself. Note that $\boldsymbol{\varphi}$ is a solution satisfying the initial condition $\boldsymbol{\varphi}(t_0) = \mathbf{x}_0$ if and only if $\boldsymbol{\varphi} = A\boldsymbol{\varphi}$.

Inspired by the contraction mapping theorem, we now consider the *successive Picard approximations* $\boldsymbol{\varphi}, A\boldsymbol{\varphi}, A^2\boldsymbol{\varphi}, \ldots$, beginning, say, with $\boldsymbol{\varphi} = \mathbf{x}_0$.

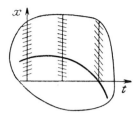

Fig. 201   An integral curve of the equation $\dot{\mathbf{x}} = \mathbf{v}(\mathbf{x}, t)$.

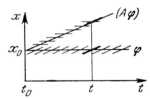

Fig. 202   The Picard mapping.

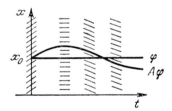

Fig. 203   Picard approximations for the equation $\dot{\mathbf{x}} = \mathbf{f}(t)$.

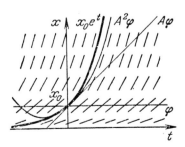

Fig. 204   Picard approximations for the equation $\dot{\mathbf{x}} = \mathbf{x}$.

Then $A^2\phi = A\phi$ too. Prf:

$$A(A\phi) = x_0 + \int_{t_0}^{t} x_0 + \int_{t_0}^{\tau} f(s)\, ds\, d\tau$$

$$= x_0 + x_0(t - t_0) + \int_{t_0}^{t} x(\tau) - x(t_0)\, d\tau$$

$$= A\phi . \checkmark$$

Example 1. Let $\dot{\mathbf{x}} = \mathbf{f}(t)$,

$$(A\varphi)(t) = \mathbf{x}_0 + \int_{t_0}^{t} \mathbf{f}(\tau)\, d\tau$$

(Fig. 203). Then the first step leads at once to an exact solution.

Example 2. Let $\dot{\mathbf{x}} = \mathbf{x}$, $t = t_0 = 0$ (Fig. 204). In this case the convergence of the approximations can written down immediately. In fact, at the point t we have

$$\varphi = \mathbf{x}_0,$$

$$A\varphi = \mathbf{x}_0 + \int_0^t \mathbf{x}_0\, d\tau = \mathbf{x}_0(1 + t),$$

$$A^2\varphi = \mathbf{x}_0 + \int_0^t \mathbf{x}_0(1 + \tau)\, d\tau = \mathbf{x}_0\left(1 + t + \frac{t^2}{2}\right),$$

$$\cdots$$

$$A^n\varphi = \mathbf{x}_0\left(1 + t + \frac{t^2}{2} + \cdots + \frac{t^n}{n!}\right),$$

$$\cdots$$

$$\lim_{n \to \infty} A^n\varphi = e^t\mathbf{x}_0.$$

Remark 1.   Thus the two definitions

1) $$e^t = \lim_{n \to \infty} \left(1 + \frac{t}{n}\right)^n,$$

2) $$e^t = 1 + t + \frac{t^2}{2!} + \cdots$$

of the exponential correspond to two ways of solving the particularly simple differential equation $\dot{x} = x$ approximately, namely the method of Euler lines and the method of sucessive Picard approximations. Historically the original definition of the exponential was simply

3) e^t is the solution of the differential equation $\dot{x} = x$ satisfying the initial condition $x(0) = 1$.

Remark 2.   The convergence of the approximations for the equation $\dot{\mathbf{x}} = k\mathbf{x}$ can be shown similarly. The reason for the convergence of the successive approximations in the general case is just that the equation $\dot{\mathbf{x}} = k\mathbf{x}$ is "the worst," i.e., the successive approximations for any equation converge no more slowly than those for some equation of the form $\dot{\mathbf{x}} = k\mathbf{x}$.

To prove the convergence of the successive approximations, we construct a complete metric space in which the Picard mapping is a contraction mapping. We begin by recalling some facts from a course on analysis.

31.2. Preliminary estimates.

1)   *The norm.* The *norm* of a vector $\mathbf{x}$ in the Euclidean space $\mathbf{R}^n$ with scalar product $(\cdot, \cdot)$ will be denoted by $|\mathbf{x}| = \sqrt{(\mathbf{x}, \mathbf{x})}$. The space $\mathbf{R}^n$ with

the metric $\rho(\mathbf{x}, \mathbf{y}) = |\mathbf{x} - \mathbf{y}|$ is a complete metric space. We note two key inequalities,† namely the *triangle inequality*

$$|\mathbf{x} + \mathbf{y}| \leqslant |\mathbf{x}| + |\mathbf{y}|$$

and the *Schwarz inequality*

$$|(\mathbf{x}, \mathbf{y})| \leqslant |\mathbf{x}|\,|\mathbf{y}|.$$

2)   *The vector integral.* Let $\mathbf{f}: [a, b] \to \mathbf{R}^n$ be a vector function with values in $\mathbf{R}^n$ which is continuous on $[a, b]$. Then the vector integral

$$\mathbf{I} = \int_a^b \mathbf{f}(t)\,dt \in \mathbf{R}^n$$

is defined in the usual way (with the help of Riemann sums).

LEMMA.   *why the outer ones?*

$$\left| \int_a^b \mathbf{f}(t)\,dt \right| \leqslant \left| \int_a^b |\mathbf{f}(t)|\,dt \right|. \tag{1}$$

Proof.   Using the triangle inequality to compare Riemann sums, we get

$$\left| \sum \mathbf{f}(t_i)\Delta_i \right| \leqslant \sum |\mathbf{f}(t_i)|\,|\Delta_i|. \quad \blacksquare$$

3)   *The norm of an operator.* Let $A: \mathbf{R}^m \to \mathbf{R}^n$ be a linear operator from one Euclidean space into another. Then we denote the norm of A by

$$|A| = \sup_{\mathbf{x} \in \mathbf{R}^n \setminus 0} \frac{|A\mathbf{x}|}{|\mathbf{x}|}.$$

Composition, not mult of tstn's.

We then have

$$|A + B| \leqslant |A| + |B|, \qquad |AB| \leqslant |A|\,|B|. \tag{2}$$

The set of linear operators from $\mathbf{R}^m$ into $\mathbf{R}^n$ becomes a complete metric space if we set $\rho(A, B) = |A - B|$.

---

† Let us recall the proof of these inequalities. Draw the two-dimensional plane through the vectors $\mathbf{x}$ and $\mathbf{y}$ of the Euclidean space. This plane inherits the Euclidean structure from $\mathbf{R}^n$. But in the Euclidean plane both inequalities are known from elementary geometry. This proves the inequalities in any Euclidean space, for example in $\mathbf{R}^n$. In particular, we have proved without any calculations at all that

$$\left| \sum_{i=1}^n x_i y_i \right|^2 \leqslant \sum_{i=1}^n x_i^2 \sum_{i=1}^n y_i^2,$$

and similarly

$$\left| \int_a^b fg\,dt \right|^2 \leqslant \int_a^b f^2\,dt \int_a^b g^2\,dt.$$

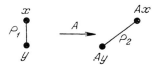

Fig. 205   The Lipschitz condition $\rho_2 \leqslant L\rho_1$.

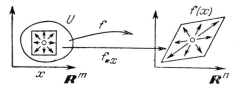

Fig. 206   The derivative of a mapping **f**.

31.3. The Lipschitz condition.

Let $A: M_1 \to M_2$ be a mapping of a metric space M_1 (with metric ρ_1) into a metric space M_2 (with metric ρ_2), and let L be a positive real number.

Definition. The mapping A is said to *satisfy a Lipschitz condition with constant L* (and we write $A \in \mathrm{Lip}\, L$) if it increases the distance between two arbitrary points of M_1 no more than L times (Fig. 205):

$$\rho_2(Ax, Ay) \leqslant L\rho_1(x, y) \quad \forall\, x, y \in M_1.$$

A mapping A is said to *satisfy a Lipschitz condition* if there exists a constant L such that $A \in \mathrm{Lip}\, L$.

Problem 1. Which of the following mappings satisfy Lipschitz conditions (the metric is Euclidean in each case):

a) $y = x^2, x \in \mathbf{R}$;   b) $y = \sqrt{x}, x > 0$;   c) $y = \sqrt{x_1^2 + x_2^2}$, $(x_1, x_2) \in \mathbf{R}^2$;

d) $y = \sqrt{x_1^2 - x_2^2}, x_1^2 \geqslant x_2^2$;   e) $y = \begin{cases} x \log x, 0 < x \leqslant 1, \\ 0, \qquad x = 0; \end{cases}$ f) $y = x^2, x \in \mathbf{C}, |x| \leqslant 1$?

Problem 2. Prove that every contraction mapping satisfies a Lipschitz condition, and that every mapping satisfying a Lipschitz condition is continuous.

31.4. Differentiability and Lipschitz conditions.

Let $\mathbf{f}: U \to \mathbf{R}^n$ be a smooth mapping (of class $C^r, r \geqslant 1$) of a domain U of the Euclidean space $\mathbf{R}^m$ into the Euclidean space $\mathbf{R}^n$ (Fig. 206). The tangent space to a Euclidean space has the natural Euclidean structure at every point, and hence the derivative

$$\mathbf{f}_*|_{\mathbf{x}} = \mathbf{f}_{*\mathbf{x}}: T\mathbf{R}_{\mathbf{x}}^m \to T\mathbf{R}_{\mathbf{f}(\mathbf{x})}^n$$

of **f** at the point $\mathbf{x} \in U \subset \mathbf{R}^m$ is a linear operator from one Euclidean space into another.

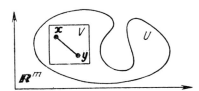

Fig. 207   Continuous differentiability implies validity of a Lipschitz condition.

THEOREM.   *Let V be any subset of the domain U which is both convex and compact. Then a continuously differentiable mapping $\mathbf{f}$ satisfies a Lipschitz condition on V with constant L equal to the least upper bound of $\mathbf{f}$ on V:*

$$L = \sup_{\mathbf{x} \in V} |\mathbf{f}_{*\mathbf{x}}|.$$

Proof.   Let $\mathbf{z}(t) = \mathbf{x} + t(\mathbf{y} - \mathbf{x})$, $0 \leqslant t \leqslant 1$ be the line segment joining the points $\mathbf{x}, \mathbf{y} \in V$ (Fig. 207). By the fundamental theorem of calculus,

$$\mathbf{f}(\mathbf{y}) - \mathbf{f}(\mathbf{x}) = \int_0^1 \frac{d}{dt} \mathbf{f}(\mathbf{z}(\tau)) \, d\tau = \int_0^1 \mathbf{f}_{*\mathbf{z}(\tau)} \dot{\mathbf{z}}(\tau) \, d\tau,$$

and hence

$$\left| \int_0^1 \mathbf{f}_{*\mathbf{z}(\tau)} \dot{\mathbf{z}}(\tau) \, d\tau \right| \leqslant \int_0^1 L |\mathbf{y} - \mathbf{x}| \, d\tau = L |\mathbf{y} - \mathbf{x}|,$$

by formulas (1) and (2), since $\dot{\mathbf{z}} = \mathbf{y} - \mathbf{x}$. ∎

Remark.   The least upper bound of the norm of the derivative $|\mathbf{f}_*|$ on V is actually achieved. In fact, $\mathbf{f} \in C^1$ by hypothesis, and hence the derivative $\mathbf{f}_*$ is continuous. It follows that $|\mathbf{f}_*|$ achieves its maximum L on the compact set V.

In undertaking the proof of the convergence of the Picard approximations, we will examine the approximations in a small neighborhood of a given point. The following four numbers will be used to describe this neighborhood.

31.5. The quantities C, L, a', b'.   Suppose the right-hand side $\mathbf{v}$ of the differential equation

$$\dot{\mathbf{x}} = \mathbf{v}(\mathbf{x}, t) \tag{3}$$

is defined and differentiable (of class C^r, $r \geqslant 1$) in a domain $U \subset \mathbf{R}^n \times \mathbf{R}^1$ of extended phase space. We fix a Euclidean structure in $\mathbf{R}^n$ and hence in $T\mathbf{R}^n_x$. Consider an arbitrary point $(\mathbf{x}_0, t_0) \in U$ (Fig. 208). The cylinder

$$\Gamma = \{\mathbf{x}, t \colon |t - t_0| \leqslant a, |\mathbf{x} - \mathbf{x}_0| \leqslant b\}$$

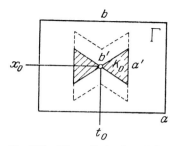

Fig. 208   The cylinder Γ and the cone K_0.

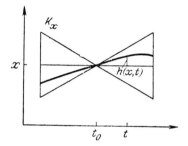

Fig. 209   Definition of $\mathbf{h}(\mathbf{x}, t)$.

lies in the domain U for sufficiently small a and b. Let C and L denote the least upper bounds of the quantities $|\mathbf{v}|$ and $|\mathbf{v}_*|$ on this cylinder, where here and subsequently the asterisk denotes the derivative (with respect to $\mathbf{x}$) for fixed t. Since the cylinder is compact, these least upper bounds are achieved:
$$|\mathbf{v}| \leqslant C, \quad |\mathbf{v}_*| \leqslant L.$$

Now let K_0 be the cone with vertex $(t_0, \mathbf{x}_0)$, "opening" C, and altitude a', so that

$$K_0 = \{x, t: |t - t_0| \leqslant a', |\mathbf{x} - \mathbf{x}_0| \leqslant C|t - t_0|\}.$$

If the number a' is small enough, the cone K_0 lies inside the cylinder Γ. Moreover, if the numbers $a', b' > 0$ are small enough, every cone K_x obtained from K_0 by parallel displacement of the vertex to the point $(t_0, \mathbf{x})$, where $|\mathbf{x} - \mathbf{x}_0| \leqslant b'$, also lies inside Γ. The numbers a' and b' are assumed to be small enough so that $K_x \subset \Gamma$, and we will look for a solution φ of equation (2) of the form $\varphi(t) = \mathbf{x} + \mathbf{h}(\mathbf{x}, t)$ subject to the initial condition $\varphi(t_0) = \mathbf{x}$ (Fig. 209). The corresponding integral curve then lies inside the cone K_x.

31.6. The metric space M.   Consider all possible continuous mappings $\mathbf{h}$ of the cylinder $|\mathbf{x} - \mathbf{x}_0| \leqslant b'$, $|t - t_0| \leqslant a'$ into the Euclidean space $\mathbf{R}^n$, and let M denote the set of such mappings which satisfy the extra condition

$$|\mathbf{h}(\mathbf{x}, t)| \leqslant C|t - t_0| \tag{4}$$

(in particular, $\mathbf{h}(\mathbf{x}, t_0) = 0$). We introduce a metric ρ in M, by setting

$$\rho(\mathbf{h}_1, \mathbf{h}_2) = \|\mathbf{h}_1 - \mathbf{h}_2\| = \max_{\substack{|\mathbf{x}-\mathbf{x}_0| \leqslant b' \\ |t-t_0| \leqslant a'}} |\mathbf{h}_1(x, t) - \mathbf{h}_2(x, t)|.$$

THEOREM.   *The set M, equipped with the metric ρ, is a complete metric space.*

Proof. A uniformly convergent sequence of continuous functions converges to a continuous function. If the functions satisfy the inequality (4) before passing to the limit, then the limit function also satisfies (4) with the same constant C.   ∎

Note that the space M depends on three positive numbers a', b', and C.

31.7. The contraction mapping $A\colon M \to M$.   Next we introduce a mapping $A\colon M \to M$ defined by†

$$(A\mathbf{h})(\mathbf{x}, t) = \int_{t_0}^{t} \mathbf{v}(\mathbf{x} + \mathbf{h}(\mathbf{x}, \tau), \tau) \, d\tau. \tag{5}$$

Because of the inequality (4), the point $(\mathbf{x} + \mathbf{h}(\mathbf{x}, \tau), \tau)$ belongs to the cone K_x, and hence to the domain of definition of the field $\mathbf{v}$.

THEOREM.   *If a' is sufficiently small, formula (5) defines a contraction mapping of the space M into itself.*

Proof.   1)   First we show that A carries M into itself. The function $A\mathbf{h}$ is continuous, since the integral of a continuous function depending continuously on a parameter is continuously dependent both on the parameter and on the upper limit. Moreover, $A\mathbf{h}$ satisfies the inequality (4), since

$$|(A\mathbf{h})(\mathbf{x}, t)| \leqslant \left| \int_{t_0}^{t} \mathbf{v}(\mathbf{x} + \mathbf{h}(\mathbf{x}, \tau), \tau) \, d\tau \right| \leqslant \left| \int_{t_0}^{t} C \, dt \right| \leqslant C|t - t_0|.$$

Therefore $AM \subset M$.

2)   Next we show that A *is a contraction mapping*, i.e., that

$$\|A\mathbf{h}_1 - A\mathbf{h}_2\| \leqslant \lambda \|\mathbf{h}_1 - \mathbf{h}_2\|, \qquad 0 < \lambda < 1.$$

To this end, we estimate $A\mathbf{h}_1 - A\mathbf{h}_2$ at the point $(\mathbf{x}, t)$. We have (Fig. 210)

$$(A\mathbf{h}_1 - A\mathbf{h}_2)(\mathbf{x}, t) = \int_{t_0}^{t} (\mathbf{v}_1 - \mathbf{v}_2) \, d\tau,$$

† In comparing (5) with the Picard mapping of Sec. 31.1, it should be borne in mind that we are now looking for a solution of the form $\mathbf{x} + \mathbf{h}$.

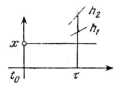

Fig. 210   Comparison of $\mathbf{v}_1$ and $\mathbf{v}_2$.

where

$$\mathbf{v}_i(\tau) = \mathbf{v}(\mathbf{x} + \mathbf{h}_i(\mathbf{x}, \tau), \tau), \qquad i = 1, 2.$$

According to Theorem 31.4, for fixed τ the function $\mathbf{v}(\mathbf{x}, \tau)$ satisfies a Lipschitz condition (in the first argument) with constant L, and hence

$$|\mathbf{v}_1(\tau) - \mathbf{v}_2(\tau)| \leqslant L|\mathbf{h}_1(\mathbf{x}, \tau) - \mathbf{h}_2(\mathbf{x}, \tau)| \leqslant L\|\mathbf{h}_1 - \mathbf{h}_2\|.$$

Moreover, according to Lemma 31.2,

$$|(A\mathbf{h}_1 - A\mathbf{h}_2)(\mathbf{x}, t)| \leqslant \left| \int_{t_0}^{t} L\|\mathbf{h}_1 - \mathbf{h}_2\| \, d\tau \right| \leqslant La'\|\mathbf{h}_1 - \mathbf{h}_2\|.$$

Therefore A is a contraction mapping if $La' < 1$.   ∎

31.8. The existence and uniqueness theorems.

COROLLARY. *Suppose the right-hand side* $\mathbf{v}$ *of the differential equation* (3) *is continuously differentiable in a neighborhood of a point* $(t_0, \mathbf{x}_0)$ *of extended phase space. Then, given any point* $\mathbf{x}$ *sufficiently close to* $\mathbf{x}_0$, *there is a neighborhood of* t_0 *in which a solution of* (3) *satisfying the initial condition* $\boldsymbol{\varphi}(t_0) = \mathbf{x}$ *is defined. Moreover, this solution depends continuously on the initial point* $\mathbf{x}$.

Proof.   According to Theorem 30.2, the contraction mapping A has a fixed point $\mathbf{h} \in M$. Let $\mathbf{g}(\mathbf{x}, t) = \mathbf{x} + \mathbf{h}(\mathbf{x}, t)$. Then

$$\mathbf{g}(\mathbf{x}, t) = \mathbf{x} + \int_{t_0}^{t} \mathbf{v}(\mathbf{g}(\mathbf{x}, \tau), \tau) \, d\tau,$$

$$\frac{\partial \mathbf{g}(x, t)}{\partial t} = \mathbf{v}(\mathbf{g}(\mathbf{x}, t), t).$$

It follows that $\mathbf{g}$ satisfies equation (3) for fixed $\mathbf{x}$ and the initial condition $\mathbf{g}(\mathbf{x}, t_0) = \mathbf{x}$ for $t = t_0$. Moreover $\mathbf{g}$ is continuous, since $\mathbf{h} \in M$.   ∎

Thus we have proved the existence theorem for equation (3) and exhibited a solution which depends continuously on the initial conditions.

Problem 1. Prove the uniqueness theorem.

Solution 1. Let $b' = 0$ in the definition of M. Then the uniqueness of the fixed point of the contraction mapping $A: M \to M$ implies the uniqueness of the solution (satisfying the initial condition $\varphi(t_0) = \mathbf{x}_0$). ∎

Solution 2. Let φ_1 and φ_2 be two solutions satisfying the same initial condition $\varphi_1(t_0) = \varphi_2(t_0) = \mathbf{x}_0$ and defined for $|t - t_0| < \alpha$. Moreover let

$$||\varphi|| = \max_{|t - t_0| < \alpha'} |\varphi(t)|,$$

where $0 < \alpha' < \alpha$. Then

$$\varphi_1(t) - \varphi_2(t) = \int_{t_0}^{t} [\mathbf{v}(\varphi_1(\tau), \tau) - \mathbf{v}(\varphi_2(\tau), \tau)] \, d\tau.$$

For sufficiently small α' the points $(\varphi_1(\tau), \tau)$ and $(\varphi_2(\tau), \tau)$ lie in the cylinder where $\mathbf{v} \in \text{Lip } L$. Therefore $||\varphi_1 - \varphi_2|| \leqslant L\alpha' ||\varphi_1 - \varphi_2||$, which implies $||\varphi_1 - \varphi_2|| = 0$ if $L\alpha' < 1$. Thus the solutions φ_1 and φ_2 coincide in some neighborhood of the point t_0. ∎

The local uniqueness theorem is now proved.

31.9. Other applications of contraction mappings.

Problem 1. Prove the inverse function theorem.

Hint. It is sufficient to invert a C^1-mapping with a unit linear part $\mathbf{y} = \mathbf{x} + \varphi(\mathbf{x})$, where $\varphi'(0) = 0$ in a neighborhood of the point $0 \in \mathbf{R}^n$ (a linear change of variables reduces the general case to this case). Suppose we look for a solution of the form $\mathbf{x} = \mathbf{y} + \psi(\mathbf{y})$. Then we get the equation

$$\psi(\mathbf{y}) = -\varphi(\mathbf{y} + \psi(\mathbf{y}))$$

for ψ. Therefore the desired function ψ is a fixed point of the mapping A defined by the formula

$$(A\psi)(\mathbf{y}) = -\varphi(\mathbf{y} + \psi(\mathbf{y})).$$

Moreover A is a contraction mapping (in a suitable metric), since the derivative of the function φ is small in a neighborhood of the point 0 (because of the condition $\varphi'(0) = 0$).

Problem 2. Prove that the Euler line approaches a solution as its step approaches zero.

Solution. Let $\mathbf{g}_\Delta = \mathbf{x} + \mathbf{h}_\Delta$ be the Euler line with step Δ and initial condition $\mathbf{g}_\Delta(\mathbf{x}, t_0) = \mathbf{x}$ (Fig. 211). In other words, let

$$\frac{\partial}{\partial t}\mathbf{g}_\Delta(\mathbf{x}, t) = \mathbf{v}(\mathbf{g}_\Delta(\mathbf{x}, s(t)), s(t)),$$

where $s(t) = t_0 + k\Delta$ and k is the integral part of $(t - t_0)/\Delta$. The difference between the

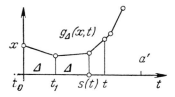

Fig. 211   The Euler line $\mathbf{g}_\Delta(\mathbf{x}, t)$.

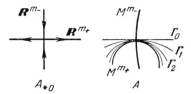

Fig. 212   Strands of the approximation A and its linear part A_{*0}.

Euler line and the solution $\mathbf{g}$ can be estimated by using the formula in Sec. 30.3:

$$\|\mathbf{g}_\Delta - \mathbf{g}\| = \|\mathbf{h}_\Delta - \mathbf{h}\| \leqslant \frac{1}{1 - \lambda}\|A\mathbf{h}_\Delta - \mathbf{h}_\Delta\|.$$

But

$$(A\mathbf{h}_\Delta)(\mathbf{x}, t) = \int_{t_0}^{t} \mathbf{v}(\mathbf{g}_\Delta(\mathbf{x}, \tau), \tau)\, d\tau,$$

$$\mathbf{h}_\Delta(\mathbf{x}, t) = \int_{t_0}^{t} \mathbf{v}(\mathbf{g}_\Delta(\mathbf{x}, s(\tau)), s(\tau))\, d\tau,$$

and as $\Delta \to 0$ the difference between the integrands approaches zero uniformly in τ, $|\tau| \leqslant a'$ (because of the equicontinuity of $\mathbf{v}$). Therefore $\|A\mathbf{h}_\Delta - \mathbf{h}_\Delta\| \to 0$ as $\Delta \to 0$, and the Euler line approaches a solution.   ∎

Problem 3. Let A be a diffeomorphism of a neighborhood of the point $0 \in \mathbf{R}^n$ onto a neighborhood of the same point carrying 0 into 0, and suppose the linear part of A at 0 (i.e., the linear operator $A_{*0}: \mathbf{R}^n \to \mathbf{R}^n$) has no eigenvalues of modulus 1. Let m_- be the number of eigenvalues with $|\lambda| < 1$ and m_+ the number of eigenvalues with $|\lambda| > 1$. Then A_{*0} has an invariant subspace $\mathbf{R}^{m_-}$ (the incoming strand) and an invariant subspace $\mathbf{R}^{m_+}$ (the outgoing strand), whose points approach 0 under application of A_{*0}^N, where $N \to +\infty$ for $\mathbf{R}^{m_-}$ and $N \to -\infty$ for $\mathbf{R}^{m_+}$ (Fig. 212).

Prove that the nonlinear mapping A also has invariant submanifolds M^{m_-} and M^{m_+} in a neighborhood of the point 0 (incoming and outgoing strands), tangent at 0 to the subspaces $\mathbf{R}^{m_-}$ and $\mathbf{R}^{m_+}$, where $A^N\mathbf{x} \to 0$ for $\mathbf{x} \in M^{m_-}$ as $N \to +\infty$ and $A^N\mathbf{x} \to 0$ for $\mathbf{x} \in M^{m_+}$ as $N \to -\infty$.

Hint. Take any submanifold Γ_0 of dimension m_+ (tangent to $\mathbf{R}^{m_+}$ at 0, say), and apply a power of A to Γ_0. Use the method of contraction mappings to prove the convergence of the resulting approximations $\Gamma_N = A^N\Gamma_0$, $N \to +\infty$ to M^{m_+}.

Problem 4. Prove the existence of incoming and outgoing strands at a nonlinear saddle point $\dot{\mathbf{x}} = \mathbf{v}(\mathbf{x})$, $\mathbf{v}(0) = 0$ (it is assumed that none of the eigenvalues of the operator $A = \mathbf{v}_*(0)$ lies on the imaginary axis).

32. The Differentiability Theorem

In this section we will eventually prove the rectification theorem.

32.1. The equation of variations.   With any differentiable mapping $\mathbf{f}: U \to V$ we can associate a linear mapping of tangent spaces at every point:

$$\mathbf{f}_{*\mathbf{x}}: TU_{\mathbf{x}} \to TV_{\mathbf{f}(\mathbf{x})}.$$

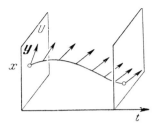

Fig. 213   Solution of the equation of variations with initial condition $(x, \mathbf{y})$.

In just the same way, we can associate with the differential equation

$$\dot{\mathbf{x}} = \mathbf{v}(\mathbf{x}, t), \qquad \mathbf{x} \in U \subset \mathbf{R}^n \tag{1}$$

a system of differential equations

$$\begin{cases} \dot{\mathbf{x}} = \mathbf{v}(\mathbf{x}, t), & \mathbf{x} \in U \subset \mathbf{R}^n, \\ \dot{\mathbf{y}} = \mathbf{v}_*(\mathbf{x}, t), & \mathbf{y} \in TU_{\mathbf{x}}, \end{cases} \tag{2}$$

linear in the tangent vector $\mathbf{y}$ (Fig. 213). We call (2) the *system of equations of variations* for equation (1). The asterisk in (2), and in subsequent formulas, denotes the derivative with respect to $\mathbf{x}$ for fixed t. Thus $\mathbf{v}_*(\mathbf{x}, t)$ is a linear operator from $\mathbf{R}^n$ into $\mathbf{R}^n$.

Together with the system (2), it is convenient to consider the system

$$\begin{cases} \dot{\mathbf{x}} = \mathbf{v}(\mathbf{x}, t), & \mathbf{x} \in U \subset \mathbf{R}^n, \\ \dot{z} = \mathbf{v}_*(\mathbf{x}, t)z, & z: \mathbf{R}^n \to \mathbf{R}^n, \end{cases} \tag{3}$$

obtained from (2) by replacing the unknown vector $\mathbf{y}$ by an unknown linear transformation z. We will apply the term *equation of variations* to the system (3) as well.

Remark.   In general, given a linear equation

$$\dot{\mathbf{y}} = A(t)\mathbf{y}, \qquad \mathbf{y} \in \mathbf{R}^n, \tag{2'}$$

it is useful to consider the associated equation

$$\dot{z} = A(t)z, \qquad z: \mathbf{R}^n \to \mathbf{R}^n, \tag{3'}$$

involving the linear operator z. From a knowledge of one of the equations (2') and (3'), we can easily find the solution of the other (how?).

32.2. The differentiability theorem.

THEOREM.   *Suppose the right-hand side* $\mathbf{v}$ *of equation* (1) *is twice continuously differentiable in a neighborhood of the point* $(\mathbf{x}_0, t_0)$. *Then the solution* $\mathbf{g}(\mathbf{x}, t)$ *of equation*

(1) *satisfying the initial condition* $\mathbf{g}(\mathbf{x}, t_0) = \mathbf{x}$ *is a continuously differentiable function of the initial condition* $\mathbf{x}$, *as* $\mathbf{x}$ *and* t *vary in some (possibly smaller) neighborhood of the point* $(\mathbf{x}_0, t_0)$:

$$\mathbf{v} \in C^2 \Rightarrow \mathbf{g} \in C_{\mathbf{x}}^1$$

(*of class* C^1 *with respect to* $\mathbf{x}$).

Proof. Since $\mathbf{v} \in C^2 \Rightarrow \mathbf{v}_* \in C^1$, the system of equations of variations (2) satisfies the conditions of Sec. 31, and the sequence of Picard approximations converges uniformly to a solution of (3) in a sufficiently small neighborhood of the point t_0. Introducing the initial conditions $\boldsymbol{\varphi}_0 = \mathbf{x}$ (sufficiently near $\mathbf{x}_0$) and $\psi_0 = E$, we denote the Picard approximations by $\boldsymbol{\varphi}_n$ (for $\mathbf{x}$) and ψ_n (for z), so that

$$\boldsymbol{\varphi}_{n+1}(\mathbf{x}, t) = \mathbf{x} + \int_{t_0}^t \mathbf{v}(\boldsymbol{\varphi}_n(\mathbf{x}, \tau), \tau)\, d\tau, \tag{4}$$

$$\psi_{n+1}(\mathbf{x}, t) = E + \int_{t_0}^t \mathbf{v}_*(\boldsymbol{\varphi}_n(\mathbf{x}, \tau), \tau)\psi_n(\mathbf{x}, \tau)\, d\tau. \tag{5}$$

Noting that $\boldsymbol{\varphi}_{0*} = \psi_0$, we deduce from (4) and (5) by induction in n that $\boldsymbol{\varphi}_{n+1*} = \psi_{n+1}$. Therefore the sequence $\{\psi_n\}$ is the sequence of derivatives of the sequence $\{\boldsymbol{\varphi}_n\}$. Both sequences (4) and (5) are uniformly convergent for sufficiently small $|t - t_0|$, being sequences of Picard approximations of the system (3). Then the sequence $\{\boldsymbol{\varphi}_n\}$ is uniformly convergent together with its derivatives with respect to $\mathbf{x}$. Hence the limit function

$$\mathbf{g}(\mathbf{x}, t) = \lim_{n \to \infty} \boldsymbol{\varphi}_n(\mathbf{x}, t)$$

is uniformly differentiable in $\mathbf{x}$.   ∎

32.3. Remark. At the same time, we have just proved the following

THEOREM. *The derivative* $\mathbf{g}_*$ *of the solution of equation* (1) *with respect to the initial condition* $\mathbf{x}$ *satisfies the equation of variations* (3) *with the initial condition* $z(t_0) = E$:

$$\frac{\partial}{\partial t}\mathbf{g}(\mathbf{x}, t) = \mathbf{v}(\mathbf{g}(\mathbf{x}, t), t),$$

$$\frac{\partial}{\partial t}\mathbf{g}_*(\mathbf{x}, t) = \mathbf{v}_*(\mathbf{g}(\mathbf{x}, t), t)\mathbf{g}_*(\mathbf{x}, t),$$

$$\mathbf{g}(\mathbf{x}, t_0) = \mathbf{x}, \quad \mathbf{g}_*(\mathbf{x}, t_0) = E.$$

This theorem explains the meaning of the equations of variations, namely they describe the action of the (t_0, t)-advance transformation on the tangent vectors to the phase space (Fig. 214).

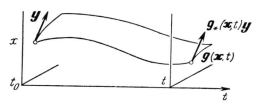

Fig. 214   Action of the (t_0, t)-advance transformation on a curve in phase space and on its tangent vector.

32.4. Higher derivatives with respect to x and t. Let $r \geqslant 2$ be an integer.

THEOREM T_r. *Suppose the right-hand side* $\mathbf{v}$ *of equation* (1) *is continuously differentiable* r *times in a neighborhood of a point* $(\mathbf{x}_0, t)$. *Then the solution* $\mathbf{g}(\mathbf{x}, t)$ *of equation* (1) *satisfying the initial condition* $\mathbf{g}(\mathbf{x}, t_0) = \mathbf{x}$ *is an* $(r - 1)$-*fold continuously differentiable function of the initial condition* $\mathbf{x}$, *as* $\mathbf{x}$ *and* t *vary in some (possibly smaller) neighborhood of the point* $(\mathbf{x}_0, t_0)$:

$$\mathbf{v} \in C^r \Rightarrow \mathbf{g} \in C_{\mathbf{x}}^{r-1}.$$

Proof. Since $\mathbf{v} \in C^r \Rightarrow \mathbf{v}_* \in C^{r-1}$, the system of equations of variations (3) satisfies the conditions of Theorem T_{r-1}. Hence Theorem T_{r-1} implies Theorem $T_r, r > 2$:

$$\mathbf{v} \in C^r \Rightarrow \mathbf{v}_* \in C^{r-1} \Rightarrow \mathbf{g}_* \in C_{\mathbf{x}}^{r-2} \Rightarrow \mathbf{g} \in C_{\mathbf{x}}^{r-1}.$$

This proves Theorem T_r, since Theorem T_2 is just Theorem 32.2. ∎

32.5. Derivatives with respect to x and t. Again let $r \geqslant 2$ be an integer.

THEOREM T_r'. *Under the conditions of Theorem* T_r, *the solution* $\mathbf{g}(\mathbf{x}, t)$ *is a differentiable function of class* C^{r-1} *with respect to both variables* $\mathbf{x}$ *and* t:

$$\mathbf{v} \in C^r \Rightarrow \mathbf{g} \in C^{r-1}.$$

The theorem is an obvious consequence of the preceding theorem. However, a formal proof goes as follows:

LEMMA. *Let* $\mathbf{f}$ *be a function (with values in* $\mathbf{R}^n$) *defined on the direct product of a domain G of Euclidean space* $\mathbf{R}^m$ *and an interval I of the t-axis:*

$$\mathbf{f}: G \times I \to \mathbf{R}^n.$$

Consider the integral

$$\mathbf{F}(\mathbf{x}, t) = \int_{t_0}^t \mathbf{f}(\mathbf{x}, \tau) \, d\tau, \qquad \mathbf{x} \in G, \quad [t_0, t] \subset I.$$

Then $\mathbf{f} \in C_{\mathbf{x}}^r$, $\mathbf{f} \in C^{r-1}$ *implies* $\mathbf{F} \in C^r$.

Proof of the lemma. Any rth partial derivative of the function $\mathbf{F}$ with respect to the variables x_i and t involving differentiation with respect to t can be expressed in terms of $\mathbf{f}$ and the partial derivatives of the function $\mathbf{f}$ of order less than r, and hence is continuous, while any rth partial derivative with respect to the variables x_i is continuous by hypothesis. ∎

Proof of the theorem. We have

$$\mathbf{g}(\mathbf{x}, t) = \mathbf{x} + \int_{t_0}^{t} \mathbf{v}(\mathbf{g}(\mathbf{x}, \tau), \tau)\, d\tau.$$

Writing $\mathbf{f}(\mathbf{x}, \tau) = \mathbf{v}(\mathbf{g}(\mathbf{x}, \tau), \tau)$ and applying the lemma, we find that

$$\mathbf{g} \in C^{p-1} \cap C_{\mathbf{x}}^{p} \Rightarrow \mathbf{g} \in C^{p}.$$

According to Theorem T_r, $\mathbf{g} \in C_{\mathbf{x}}^{p}$ for $p < r$. Thus we get successively

$$\mathbf{g} \in C^{0} \Rightarrow \mathbf{g} \in C^{1} \Rightarrow \cdots \Rightarrow \mathbf{g} \in C^{r-1}.$$

But $\mathbf{g} \in C^{0}$ by Sec. 31.8 (the solution depends continuously on $\mathbf{x}$, t). This completes the proof of Theorem T_r'. ∎

Problem 1. Prove that if the right-hand side of the differential equation (1) is infinitely differentiable, then the solution is also an infinitely differentiable function of the initial conditions:

$$\mathbf{v} \in C^{\infty} \Rightarrow \mathbf{g} \in C^{\infty}.$$

Remark. It can also be shown that if the right-hand side $\mathbf{v}$ is analytic (has a Taylor series converging to $\mathbf{v}$ in a neighborhood of every point), then the solution $\mathbf{g}$ also depends analytically on $\mathbf{x}$ and t. It is natural to study differential equations with analytic right-hand sides for both complex values of the unknowns and (of particular importance) for complex values of the time t.[†]

32.6. The rectification theorem. This theorem is an obvious consequence of Theorem T_r'. Before proving it, we recall two simple geometric propositions. Let L_1 and L_2 be two linear subspaces of a third linear space L (Fig. 215). Then L_1 and L_2 are said to be *transverse* if their sum is the whole space $L\colon L_1 + L_2 = L$. For example, a line in $\mathbf{R}^3$ is transverse to a plane if it intersects the plane at a nonzero angle.

PROPOSITION 1. *Every k-dimensional subspace $\mathbf{R}^k$ in $\mathbf{R}^n$ has an $(n - k)$-dimensional transverse subspace (in fact, at least one of the C_k^n coordinate planes of the space $\mathbf{R}^{n-k}$ will be transverse to $\mathbf{R}^k$).*

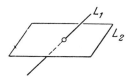

Fig. 215   The line L_1 is transverse to the plane L_2 in the space $\mathbf{R}^3$.

[†] Concerning this theory, see e.g., V. V. Golubev, *Lectures on the Analytic Theory of Differential Equations* (in Russian), Moscow (1950).

The proof is given in courses on linear algebra (the theorem on the rank of a matrix).

PROPOSITION 2. *If a linear mapping $A: L \to M$ maps any two transverse subspaces onto transverse subspaces, then it maps L onto the whole space M.*

Proof. $AL = AL_1 + AL_2 = M$. ∎

Proof of the rectification theorem: *Nonautonomous case* (see Sec. 8.1). Consider the mapping G of a domain of the direct product $\mathbf{R}^n \times \mathbf{R}$ into the extended phase space of the equation

$$\dot{\mathbf{x}} = \mathbf{v}(\mathbf{x}, t) \tag{1}$$

defined by the formula $G(\mathbf{x}, t) = (\mathbf{g}(\mathbf{x}, t), t)$, where $\mathbf{g}(\mathbf{x}, t)$ is a solution of (1) satisfying the initial condition $\mathbf{g}(\mathbf{x}, t_0) = \mathbf{x}$. Then, as we now show, G is a rectifying diffeomorphism in a neighborhood of the point $(\mathbf{x}_0, t_0)$.

   a) *The mapping G is differentiable* (of class C^{r-1} if $\mathbf{v} \in C^r$), by Theorem T_r'.

   b) *The mapping G leaves t unchanged*: $G(x, t) = (\mathbf{g}(\mathbf{x}, t), t)$.

   c) *The mapping G_* carries the standard vector field* $\mathbf{e}$ ($\dot{\mathbf{x}} = 0$, $t = 1$) *into the given field*, i.e., $G_* \mathbf{e} = (\mathbf{v}, 1)$, since $\mathbf{g}(\mathbf{x}, t)$ is a solution of (1).

   d) *The mapping G is a diffeomorphism in a neighborhood of the point $(\mathbf{x}_0, t_0)$.* In fact, calculating the restriction of the linear operator $G_*|_{t_0, \mathbf{x}_0}$ to the transverse planes $\mathbf{R}^n$ and $\mathbf{R}^1$ (Fig. 216), we get

$$G_*|_{\mathbf{R}^n : t = t_0} = E, \qquad G_*|_{\mathbf{R}^1 : x = x_0} \mathbf{e} = \mathbf{v} + \mathbf{e}.$$

The plane $\mathbf{R}^n$ and the line with direction $\mathbf{v} + \mathbf{e}$ are transverse. Therefore G_* is a linear mapping of $\mathbf{R}^{n+1}$ onto $\mathbf{R}^{n+1}$ and hence an isomorphism (the Jacobian of G_* at the point $(\mathbf{x}_0, t_0)$ is nonzero). It follows from the inverse function theorem that G is a local diffeomorphism.

Proof of the rectification theorem: *Autonomous case* (see Sec. 7.1). Consider the autonomous equation

$$\dot{\mathbf{x}} = \mathbf{v}(\mathbf{x}), \qquad \mathbf{x} \in U \subset \mathbf{R}^n. \tag{6}$$

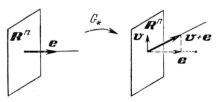

Fig. 216   Derivative of the mapping G at the point $(\mathbf{x}_0, t_0)$.

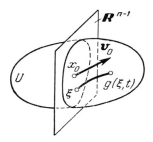

Fig. 217   Construction of the diffeomorphism rectifying a vector field.

Let the phase velocity $\mathbf{v}_0$ at the point $\mathbf{x}_0$ be different from 0 (Fig. 217). Then there exists an $(n - 1)$-dimensional hyperplane $\mathbf{R}^{n-1} \subset \mathbf{R}^n$ passing through $\mathbf{x}_0$ and transverse to $\mathbf{v}_0$ (more exactly, a corresponding plane in the tangent space $TU_{\mathbf{x}_0}$ transverse to the line $\mathbf{R}^1$ with direction $\mathbf{v}_0$). Let G be the mapping of the domain $\mathbf{R}^{n-1} \times \mathbf{R}$ where $\mathbf{R}^{n-1} = \{\xi\}$, $\mathbf{R} = \{t\}$ into the domain $\mathbf{R}^n$ defined by the formula $G(\xi, t) = \mathbf{g}(\xi, t)$ where ξ lies on $\mathbf{R}^{n-1}$ near $\mathbf{x}_0$ and $\mathbf{g}(\xi, t)$ is the value of the solution of equation (6) satisfying the initial condition $\boldsymbol{\varphi}(0) = \xi$ at the time t. Then, as we now show, G^{-1} is a rectifying diffeomorphism in a sufficiently small neighborhood of the point $(\xi = \mathbf{x}_0, t = 0)$.

a) *The mapping G is differentiable* ($G \in C^{r-1}$ if $\mathbf{v} \in C^r$), by Theorem T'_r.

b) *The mapping G^{-1} is rectifying*, since G_* carries the standard vector field $\mathbf{e}$ ($\dot{\xi} = 0, \dot{t} = 1$) into $G_*\mathbf{e} = \mathbf{v}$, because $\mathbf{g}(\xi, t)$ satisfies equation (6).

c) *The mapping G is a local diffeomorphism.* In fact, calculating the linear operator $G_*|_{\mathbf{x}_0, t_0}$ on the transverse planes $\mathbf{R}^{n-1}$ and $\mathbf{R}^1$, we get

$$G_*|_{\mathbf{R}^{n-1}} = E, \qquad G_*|_{\mathbf{R}^1} \mathbf{e} = \mathbf{v}_0.$$

Thus the operator $G_*|_{\mathbf{x}_0, t_0}$ carries the pair of transverse subspaces $\mathbf{R}^{n-1}$ and $\mathbf{R}^1 \subset \mathbf{R}^n$ into a pair of transverse subspaces. Therefore $G_*|_{\mathbf{x}_0, t_0}$ is a linear mapping of $\mathbf{R}^n$ onto $\mathbf{R}^n$, and hence an isomorphism. It follows from the inverse function theorem that G is a local diffeomorphism ($f = G^{-1}$ in the notation of Sec. 7).

Remark. Since the differentiation theorem was proved with the loss of one derivative ($\mathbf{v} \in C^r \Rightarrow \mathbf{g} \in C^{r-1}$), we can only guarantee that the rectifying diffeomorphisms belong to the smoothness class C^{r-1}. However, as will be shown below, the diffeomorphism just constructed is actually of class C^r.

32.7. The last derivative. In proving Theorem 32.2, the field $\mathbf{v}$ was assumed to be twice continuously differentiable, but actually mere continuous differentiability is sufficient.

THEOREM. *If the right-hand side* $\mathbf{v}(\mathbf{x}, t)$ *of the differential equation* $\dot{\mathbf{x}} = \mathbf{v}(\mathbf{x}, t)$ *is continuously differentiable, then the solution* $\mathbf{g}(\mathbf{x}, t)$ *satisfying the initial condition* $\mathbf{g}(\mathbf{x}, t_0) = \mathbf{x}$ *is a continuously differentiable function of the initial condition:*

$$\mathbf{v} \in C^1 \Rightarrow \mathbf{g} \in C_{\mathbf{x}}^1. \tag{7}$$

COROLLARIES.

1) $\mathbf{v} \in C^r \Rightarrow \mathbf{g} \in C^r$ *for* $r \geqslant 1$.

2) *If* $\mathbf{v} \in C^r$, *the rectifying diffeomorphism constructed in Sec. 32.6 is continuously differentiable* r *times.*

The corollaries are deduced from (7) by literal repetition of the considerations of Secs. 32.4–32.6. However, the proof of the theorem (7) itself requires some ingenuity.

Proof of the theorem. We begin with the following considerations.

LEMMA 1. *Let*

$$\dot{\mathbf{y}} = A(t)\mathbf{y}, \tag{8}$$

be a linear equation whose right-hand side is a continuous function of t. *Then the solution of* (8) *exists continuously, is uniquely determined by the initial condition* $\boldsymbol{\varphi}(t_0) = \mathbf{y}_0$, *and depends continuously on* $\mathbf{y}_0$ *and* t.

Proof. The proof of the existence, uniqueness, and continuity theorems (Sec. 31) uses only the differentiability with respect to $\mathbf{x}$ for fixed t (actually only the existence of a Lipschitz condition in $\mathbf{x}$). Therefore the proof remains valid if the dependence on t is assumed to be only continuous. ∎

Note that the solution depends linearly on $\mathbf{y}_0$ and is a continuously differentiable function of t. Therefore the solution belongs to the class C^1 with respect to both $\mathbf{y}_0$ and t.

LEMMA 2. *If the linear operator* A *in Lemma 1 also depends on a parameter* α *and if the function* $A(t, \alpha)$ *is continuous, then the solution is a continuous function of* $\mathbf{y}_0$, t, *and* α.

Proof. The solution can be constructed as the limit of a sequence of Picard approximations, where each approximation is a continuous function of $\mathbf{y}_0$, t, and α. The sequence of approximations is uniformly convergent in the variables $\mathbf{y}_0$, t, and α, as the latter vary in a sufficiently small neighborhood of any point $(\mathbf{y}_0, t_0, \alpha_0)$. Hence the limit is a continuous function of $\mathbf{y}_0$, t, and α. ∎

We now apply Lemma 2 to the equation of variations.

LEMMA 3. *The system of equations of variations*

$$\begin{cases} \dot{\mathbf{x}} = \mathbf{v}(\mathbf{x}, t), \\ \dot{\mathbf{y}} = \mathbf{v}_*(\mathbf{x}, t)\mathbf{y} \end{cases}$$

has a solution which is uniquely determined by its initial conditions and depends continuously on these conditions, provided only that the field $\mathbf{v}$ *is of class* C^1.

Proof. By the existence theorem of Sec. 31.8, the first equation of the system has a solution, which is uniquely determined by its initial conditions $\mathbf{x}_0$, t_0 and depends on these conditions continuously. Substituting this solution into the second equation, we get a linear equation in $\mathbf{y}$ whose right-hand side depends continuously on t and also on the initial condition $\mathbf{x}_0$ (regarded as a parameter) of the solution of the first equation. But, by Lemma 2, this linear equation has a solution which is determined by its initial data $\mathbf{y}_0$ and is a continuous function of t, $\mathbf{y}_0$, and the parameter $\mathbf{x}_0$. ∎

Thus *the equation of variations is solvable even in the case* $\mathbf{v} \in C^1$. Note that in the case $\mathbf{v} \in C^2$ we proved that the derivative of the solution with respect to the initial data satisfies the equation of variations (3), but this can no longer be asserted, since we still do not know whether this derivative exists.

To prove the differentiability of the solution with respect to the initial conditions, we first consider a special case.

LEMMA 4. *Suppose the vector field* $\mathbf{v}(\mathbf{x}, t)$ *of class* C^1 *and its derivative* $\mathbf{v}_*$ *both vanish at the point* $\mathbf{x} = 0$ *for all* t. *Then the solution of the equation* $\dot{\mathbf{x}} = \mathbf{v}(\mathbf{x}, t)$ *is differentiable with respect to the initial conditions at the point* $\mathbf{x} = 0$.

Proof. By hypothesis,

$$|\mathbf{v}(\mathbf{x}, t)| = o(|\mathbf{x}|)$$

in a neighborhood of the point $\mathbf{x} = 0$. Using the formula of Sec. 30.3 to estimate the error of the approximation $\mathbf{x} = \mathbf{x}_0$ to the solution $\mathbf{x} = \boldsymbol{\varphi}(t)$ satisfying the initial condition $\boldsymbol{\varphi}(t_0) = \mathbf{x}_0$, we find that

$$|\boldsymbol{\varphi} - \mathbf{x}_0| \leqslant \frac{1}{1 - \lambda} \left| \int_{t_0}^{t} \mathbf{v}(\mathbf{x}_0, \tau) \, d\tau \right| \leqslant K \max_{t_0 \leqslant \tau \leqslant t} |\mathbf{v}(\mathbf{x}_0, \tau)|$$

for sufficiently small $|\mathbf{x}_0|$ and $|t - t_0|$, where the constant K is independent of $\mathbf{x}_0$. Thus $|\boldsymbol{\varphi} - \mathbf{x}_0| = o(|\mathbf{x}_0|)$, which implies that $\boldsymbol{\varphi}$ is differentiable with respect to $\mathbf{x}_0$ at zero. ∎

We now reduce the general case to the special situation of Lemma 4. To do so, we need only choose a suitable coordinate system in extended phase space. First we note that the solution under consideration can always be regarded as the null solution:

LEMMA 5. *Let* $\mathbf{x} = \boldsymbol{\varphi}(t)$ *be a solution of the equation* $\dot{\mathbf{x}} = \mathbf{v}(\mathbf{x}, t)$ *with a right-hand side of class* C^1, *defined in a domain of extended phase space* $\mathbf{R}^n \times \mathbf{R}^1$. *Then there exists a* C^1-*diffeomorphism of extended phase space which preserves time, i.e.,* $(\mathbf{x}, t) \rightarrow (\mathbf{x}_1(\mathbf{x}, t), t)$, *and carries the solution* $\boldsymbol{\varphi}$ *into* $\mathbf{x}_1 \equiv 0$.

Proof. Since $\boldsymbol{\varphi} \in C^1$, we need only make the shift $\mathbf{x}_1 = \mathbf{x} - \boldsymbol{\varphi}(t)$. ∎

In the system of coordinates $(\mathbf{x}_1, t)$, the right-hand side of our equation equals 0 at the point $\mathbf{x}_1 = 0$. We now show that the derivative of the right-hand side with respect to $\mathbf{x}_1$ can also be made to vanish with the help of a suitable change of coordinates which is linear in $\mathbf{x}$.

LEMMA 6. *Under the conditions of Lemma 5, the coordinates* $(\mathbf{x}_1, t)$ *can be chosen in such a way that the equation* $\dot{\mathbf{x}} = \mathbf{v}(\mathbf{x}, t)$ *is equivalent to the equation* $\dot{\mathbf{x}}_1 = \mathbf{v}_1(\mathbf{x}_1, t)$, *where the field* $\mathbf{v}_1$ *and its derivative* $\partial \mathbf{v}_1 / \partial \mathbf{x}_1$ *both vanish at the point* $\mathbf{x}_1 = 0$. *Moreover, the function* $\mathbf{x}_1(\mathbf{x}, t)$ *can be chosen to be linear (but not necessarily homogeneous) in* $\mathbf{x}$.

According to Lemma 5, we can assume that $\mathbf{v}_1(0, t) = 0$.

To prove Lemma 6, we first consider the following special case:

LEMMA 7. *The assertion of Lemma 6 is valid for the linear equation* $\dot{\mathbf{x}} = A(t)\mathbf{x}$.

Proof. We need only choose $\mathbf{x}_1$ to be the value of the solution satisfying the initial condition $\boldsymbol{\varphi}(t) = \mathbf{x}$ at a fixed time t_0. According to Lemma 1, $\mathbf{x}_1 = B(t)\mathbf{x}$ where $B(t): \mathbf{R}^n \rightarrow \mathbf{R}^n$ is a linear operator of class C^1 in t. But our linear equation takes the form $\dot{\mathbf{x}}_1 = 0$ in the coordinates $(\mathbf{x}_1, t)$. ∎

Proof of Lemma 6. First we linearize the equation $\dot{\mathbf{x}} = \mathbf{v}(\mathbf{x}, t)$ at zero, i.e., we form the equation of variations

$$\dot{\mathbf{x}} = A(t)\mathbf{x}, \qquad A(t) = \mathbf{v}_*(0, t).$$

By hypothesis $\mathbf{v} \in C^1$, and hence $A \in C^0$. By Lemma 7, we can choose C^1-coordinates $\mathbf{x}_1 = B(t)\mathbf{x}$ such that the linearized equation is of the form $\dot{\mathbf{x}}_1 = 0$ in the new coordinates. It is easy to see that the right-hand side of the original nonlinear equation has a zero linear part in this coordinate system. In fact, let $\mathbf{v} = A\mathbf{x} + \mathbf{Q}$, $\mathbf{x} = C\mathbf{x}_1$ (so that $\mathbf{Q} = o(|\mathbf{x}|)$, $C = B^{-1}$). Making these substitutions in the equation $\dot{\mathbf{x}} = \mathbf{v}$, we get the differential equation for $\mathbf{x}_1$:

$$\dot{C}\mathbf{x}_1 + C\dot{\mathbf{x}}_1 = AC\mathbf{x}_1 + \mathbf{Q}.$$

But, by the definition of C, the first terms on the left and right (the terms linear in $\mathbf{x}_1$) are equal, and hence

$$\dot{\mathbf{x}}_1 = C^{-1}\mathbf{Q}(C\mathbf{x}_1, t) = o(|\mathbf{x}_1|). \quad \blacksquare$$

Combining Lemmas 6 and 4, we deduce

LEMMA 8. *The solution of the differential equation $\dot{\mathbf{x}} = \mathbf{v}(\mathbf{x}, t)$ with a right-hand side of class C^1 depends differentiably on the initial condition. The derivative z of the solution with respect to the initial condition satisfies the system of equations of variations*

$$\dot{\mathbf{x}} = \mathbf{v}(\mathbf{x}, t), \qquad \dot{z} = \mathbf{v}_*(\mathbf{x}, t)z, \qquad z(t_0) = E \colon \mathbf{R}^n \to \mathbf{R}^n.$$

Proof. Write the equation in the coordinate system of Lemma 6 and then apply Lemma 4. $\blacksquare$

To prove the theorem, we now need only verify the continuity of the derivative of the solution with respect to the initial condition. According to Lemma 8, this derivative exists and satisfies the system of equations of variations. It follows from Lemma 3 that the solutions of this system depend continuously on $\mathbf{x}_0$ and t, and the theorem is finally proved. $\blacksquare$

5 Differential Equations on Manifolds

In this chapter we define differentiable manifolds, proving a theorem on existence of a phase flow determined by a vector field on a manifold. Lack of space will not permit us to go into the many interesting and deep results that have been obtained in the theory of differential equations on manifolds. The present chapter is intended merely as an introduction to this subject, which lies at the junction of analysis and topology.

33. Differentiable Manifolds

The concept of a differentiable or smooth manifold plays just as fundamental a role in geometry and analysis as the concepts of group and linear space play in algebra.

33.1. Examples of manifolds. Once manifolds are defined (below), we will find that the following objects are all manifolds (Fig. 218):

1) The linear space $\mathbf{R}^n$ or any domain (open subset) U of $\mathbf{R}^n$.

2) The sphere S^n defined by the equation $x_1^2 + \cdots + x_{n+1}^2 = 1$ in the Euclidean space $\mathbf{R}^{n+1}$, in particular, the circle S^1.

3) The torus $T^2 = S^1 \times S^1$ (cf. Sec. 24).

4) The projective space

$$\mathbf{RP}^n = \{(x_0 : x_1 : \cdots : x_n)\}.$$

It will be recalled that the points of this space are straight lines passing through the origin of coordinates in $\mathbf{R}^{n+1}$. Such a line is specified by any of its points (other than 0). The coordinates of this point $(x_0, x_1, \ldots, x_n)$ in $\mathbf{R}^{n+1}$ are called the *homogeneous coordinates* of the corresponding point of projective space.

The last example is particularly useful. In considering the definitions that follow, it will be useful to think in terms of affine coordinates in a projective space (see Sec. 33.3, Example 3).

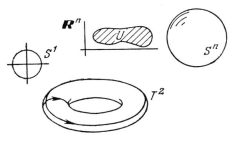

Fig. 218 Examples of manifolds.

33.2. Definitions. A *differentiable manifold* M is a set M equipped with a differentiable structure. To equip M with a *differentiable structure* or *manifold structure*, we specify an *atlas* consisting of *maps* which are *compatible*.

Definition 1. By a *map* is meant a domain $U \subset \mathbf{R}^n$ together with a one-to-one mapping $\varphi\colon W \to U$ of a subset W of the set M onto U (Fig. 219). We call $\varphi(x)$ an *image* of the point $x \in W \subset M$ on the map U.

Consider two maps

$$\varphi_i\colon W_i \to U_i, \qquad \varphi_j\colon W_j \to U_j$$

(Fig. 220). If the sets W_i and W_j intersect, then their intersection $W_i \cap W_j$ has an image on both maps:

$$U_{ij} = \varphi_i(W_i \cap W_j), \qquad U_{ji} = \varphi_j(W_j \cap W_i).$$

The transformation from one map to another is specified by the following mapping between *subsets of linear spaces*:

$$\varphi_{ij}\colon U_{ij} \to U_{ji}, \qquad \varphi_{ij}(x) = \varphi_j(\varphi_i^{-1}(x)).$$

Definition 2. Two maps

$$\varphi_i\colon W_i \to U_i, \qquad \varphi_j\colon W_j \to U_j$$

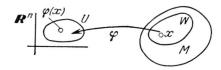

Fig. 219   A map.

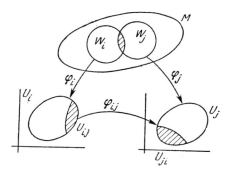

Fig. 220   Compatible maps.

are called *compatible* if
1) The sets U_{ij} and U_{ji} are open (possibly empty);
2) The mappings φ_{ij} and φ_{ji} (defined if $W_i \cap W_j$ is nonempty) are diffeomorphisms of domains of $\mathbf{R}^n$.

Remark. Depending on the smoothness class of the mappings φ_{ij}, we get different classes of manifolds. If by a diffeomorphism we mean a diffeomorphism of class C^r, $i \leqslant r \leqslant \infty$, then the manifold (defined by the atlas giving rise to the mappings φ_{ij}) will be called a *differentiable manifold of class* C^r. If $r = 0$, so that the φ_{ij} are only required to be homeomorphisms, we get the definition of a *topological manifold*. If we require the φ_{ij} to be analytic,† we get *analytic manifolds*.

There are other possibilities as well. For example, fixing an orientation in $\mathbf{R}^n$ and requiring that the diffeomorphisms φ_{ij} preserve this orientation (i.e., that the φ_{ij} have positive Jacobians at every point), we arrive at the definition of an *oriented manifold*.

Definition 3. By an *atlas* on M is meant a set of maps $\varphi_i: W_i \to U_i$ such that
1) Every pair of maps is compatible;
2) Every point $x \in M$ has an image on at least one map.

Definition 4. Two atlases on M are said to be *equivalent* if their union is itself an atlas (i.e., if every map of the first atlas is compatible with every map of the second).

It is easy to see that Definition 4 actually defines an equivalence relation.

Definition 5. By a *differentiable structure* on M is meant a class of equivalent atlases.

At this point, we note two conditions often imposed on manifolds to avoid pathology:

1) *Separability.* Any two points $x, y \in M$ have nonintersecting neighborhoods (Fig. 221), i.e., either there exist two maps

$$\varphi_i: W_i \to U_i, \qquad \varphi_j: W_j \to U_j$$

with nonintersecting W_i and W_j containing x and y respectively, or there exists a map on which both points x and y have images.

If separability is not required, then the set obtained from two lines $\mathbf{R} = \{x\}$, $\mathbf{R} = \{y\}$ by identifying points with equal negative coordinates x and y will be a manifold. The theorem on unique extension of solutions of differential equations will fail to hold on such manifolds, although the local uniqueness theorem will be true.

† A function is said to be analytic if it is the sum of its own Taylor series in a neighborhood of every point.

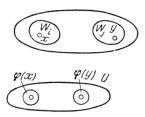

Fig. 221   Separability.

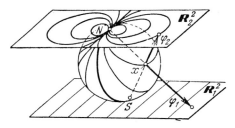

Fig. 222   Atlas of a sphere. The family of circles on the sphere tangent at the point N is represented on the lower map by a family of parallel lines and on the upper map by a family of tangent circles.

2) *Countability*. There exists an atlas M with no more than a countable number of maps.

Henceforth the term "manifold" will mean a differentiable manifold satisfying the separability and countability conditions.

33.3. Examples of atlases.

1) The sphere S^2 with equation $x_1^2 + x_2^2 + x_3^2 = 1$ in $\mathbf{R}^3$ can be equipped with an atlas consisting of two maps, for example by using stereographic projection (Fig. 222). Here we have

$$W_1 = S^2 \setminus N, \qquad U_1 = \mathbf{R}_1^2,$$
$$W_2 = S^2 \setminus S, \qquad U_2 = \mathbf{R}_2^2.$$

Problem 1. Write formulas for the mappings $\varphi_{1,2}$ and verify that the two maps are compatible.

Similarly, we can use an atlas consisting of two maps to define a differentiable structure in S^n.

2) An atlas for the torus can be constructed by using angular coordinates, namely the latitude θ and the longitude ψ (Fig. 223). For example, we can

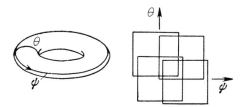

Fig. 223   Atlas of a torus.

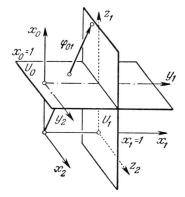

Fig. 224   Affine maps of the projective plane.

consider the four maps obtained when θ and ψ vary in the intervals

$$0 < \theta < 2\pi, \qquad -\pi < \theta < \pi,$$
$$0 < \psi < 2\pi, \qquad -\pi < \psi < \pi.$$

3) An atlas for the projective plane $\mathbf{RP}^2$ can be made up of the following three "affine maps" (Fig. 224):

$$x_0 : x_1 : x_2 \begin{cases} \xrightarrow{\ \varphi_0\ } y_1 = \dfrac{x_1}{x_0}, \quad y_2 = \dfrac{x_2}{x_0} & \text{if } x_0 \neq 0, \\[2mm] \xrightarrow{\ \varphi_1\ } z_1 = \dfrac{x_0}{x_1}, \quad z_2 = \dfrac{x_2}{x_1} & \text{if } x_1 \neq 0, \\[2mm] \xrightarrow{\ \varphi_2\ } u_1 = \dfrac{x_0}{x_2}, \quad u_2 = \dfrac{x_1}{x_2} & \text{if } x_2 \neq 0. \end{cases}$$

These maps are compatible. For example, compatibility of φ_0 and φ_1 means that the mapping $\varphi_{0,1}$ of the domain $U_{0,1} = \{y_1, y_2 : y_1 \neq 0\}$ of the plane (y_1, y_2) onto the domain $U_{1,0} = \{z_1, z_2 : z_1 \neq 0\}$ of the plane

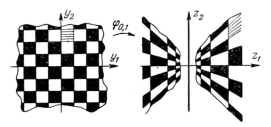

Fig. 225   Compatibility of maps of the projective plane.

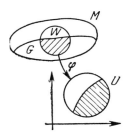

Fig. 226   An open subset.

(z_1, z_2) given by the formulas $z_1 = y_1^{-1}$, $z_2 = y_2 y_1^{-1}$ is a diffeomorphism (Fig. 225).

Proof. Note that $y_1 = z_1^{-1}, y_2 = z_2 z_1^{-1}$.   ∎

Similarly, we can use an atlas consisting of $n + 1$ affine maps to equip the projective space $\mathbf{RP}^n$ with a differentiable structure.

33.4. Compactness.

Definition. A subset G of a manifold M is said to be *open* if its image $\varphi(W \cap G)$ on every map $\varphi : W \to U$ is an open subset of the domain U of linear space (Fig. 226).

Problem 1. Prove that the intersection of two and the union of any number of open subsets of a manifold is open.

Definition. A subset K of a manifold M is said to be *compact* if every covering of the set K by open sets has a finite subcovering.

Problem 2. Prove that the sphere S^n is compact. Is the projective space $\mathbf{RP}^n$ compact?

Hint. Use the following theorem.

THEOREM. *Suppose a subset F of a manifold M (Fig. 227) is the union of a finite*

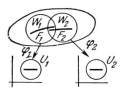

Fig. 227   A compact subset.

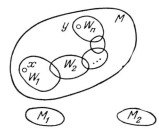

Fig. 228   A connected manifold M and a disconnected manifold $M_1 \cup M_2$.

number of subsets F_i, each of which has a compact image on one of the maps $F_i \subset W_i$, $\varphi_i \colon W_i \to U_i$, where $\varphi_i(F_i)$ is a compact set in $\mathbf{R}^n$. Then F is compact.

Proof. Let $\{G_j\}$ be an open covering of the set F. Then $\{\varphi_i(G_j \cap W_i)\}$ is an open covering of the compact set $\varphi_i(F_i)$ for every i. Letting j range over the resulting finite set of values, we get a finite number of G_j covering F.   ∎

33.5. Connectedness and dimension.

Definition. A manifold M is said to be *connected* (Fig. 228) if given any two points $x, y \in M$, there exists a finite chain of maps $\varphi_i \colon W_i \to U_i$ such that W_1 contains x, W_n contains y, $W_i \cap W_{i+1}$ $\forall i$ is nonempty, and U_i is connected.

A disconnected manifold M decomposes into *connected components* M_i.†

Problem 1. Are the manifolds defined by

$$x^2 + y^2 - z^2 = C, \qquad C \neq 0$$

in $\mathbf{R}^3$ (in $\mathbf{RP}^3$) connected?

Problem 2. The set of all matrices of order n with nonzero determinants has the natural structure of a differentiable manifold (a domain in $\mathbf{R}^{n^2}$). How many connected components does this manifold have?

THEOREM. *Let M be a connected manifold, and let*

$$\varphi_i \colon W_i \to U_i$$

† I.e., any two points of U_i can be joined by a polygonal curve in $U_i \subset \mathbf{R}^n$.

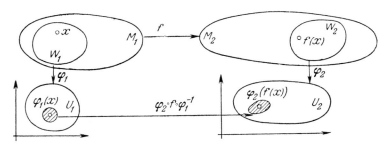

Fig. 229 A differentiable mapping.

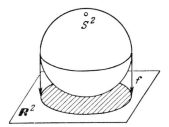

Fig. 230 Projection of a sphere onto the plane gives a closed disk.

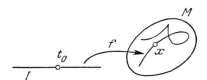

Fig. 231 A curve on a manifold M.

be its maps. Then all the linear spaces $\mathbf{R}^n$ containing the domains U_i have the same dimension.

Proof. A consequence of the fact that a diffeomorphism between domains of linear spaces is possible only if the spaces have the same dimension and the fact that any two domains W_i and W_j of a connected manifold M can be joined by a finite chain of pairwise intersecting domains. ∎

The number n figuring in the theorem is called the *dimension* of the manifold M, denoted by $\dim M$. For example,

$$\dim \mathbf{R}^n = \dim S^n = \dim T^n = \dim \mathbf{RP}^n = n.$$

A disconnected manifold is said to be n-dimensional if all its connected components have the same dimension n.

Problem 3. Equip the set $O(n)$ of all orthogonal matrices of order n with the structure of a differentiable manifold. Find its connected components and its dimension.

Ans. $O(n) = SO(n) \times \mathbf{Z}_2$, $\dim O(n) = \dfrac{n(n-1)}{2}$.

33.6. Differentiable mappings.

Definition. A mapping $f: M_1 \to M_2$ of one C^r-manifold into another is said to be *differentiable* (of class C^r) if it is given by differentiable functions (of class C^r) in the local coordinates on M_1 and M_2.

In other words, let $\varphi_1: W_1 \to U_1$ be a map of M_1, acting on a neighborhood of a point $x \in M_1$, and $\varphi_2: W_2 \to U_2$ a map of M_2, acting on a neighborhood of a point $f(x) \in W_2$ (Fig. 229). Then the mapping of domains of Euclidean space $\varphi_2 \circ f \circ \varphi_1^{-1}$ defined in a neighborhood of the point $\varphi_1(x)$ must be differentiable of class C^r.

Example 1. The projection of a sphere onto the plane (Fig. 230) is a differentiable mapping. Note that a differentiable mapping need not carry a differentiable manifold into a differentiable manifold

Example 2. By a *curve*† on a manifold M leaving the point $x \in M$ at time t_0 is meant a differentiable mapping $f: I \to M$ of an interval I of the real t-axis containing the point t_0 into a manifold M such that $f(t_0) = x$.

Example 3. By a *diffeomorphism* $f: M_1 \to M_2$ of a manifold M_1 onto a manifold M_2 is meant a differentiable mapping f, whose inverse $f^{-1}: M_2 \to M_1$ exists and is differentiable. Two manifolds M_1 and M_2 are said to be *diffeomorphic* if there exists a diffeomorphism from one onto the other. For example, the sphere and the ellipsoid are diffeomorphic.

33.7. Remark.
It is easy to see that every connected one-dimensional manifold is diffeomorphic to a circle (if it is compact) or to a line (if it is noncompact).

Examples of two-dimensional manifolds are the sphere, the torus (diffeomorphic to a "sphere with one handle") and the "sphere with n handles" (Fig. 232).

Fig. 232   Nondiffeomorphic two-dimensional manifolds.

† Synonymously, a *parametrized curve*, since one-dimensional submanifolds (defined in Sec 33.8) of the manifold M are sometimes also called curves on M. A parametrized curve can have points of self-intersection, cusps, etc. (Fig. 231).

In courses on topology it is proved that every two-dimensional compact connected oriented manifold is diffeomorphic to a sphere with $n \geqslant 0$ handles. Little is known about three-dimensional manifolds. For example, it is not known whether every compact simply connected[†] three-dimensional manifold is diffeomorphic to the sphere S^3 (*Poincaré's hypothesis*) or even homeomorphic to S^3.

The differentiable and topological classifications of manifolds do not coincide in higher dimensions. For example, there exist precisely 28 smooth manifolds, called *Milnor spheres*, homeomorphic to the sphere S^7, but not diffeomorphic to each other.

A Milnor sphere in $\mathbf{C}^5$ with coordinates $z_1, \ldots, z_5$ is defined by the two equations

$$z_1^{6k-1} + z_2^3 + z_3^2 + z_4^2 + z_5^2 = 0,$$
$$|z_1|^2 + \cdots + |z_5|^2 = 1.$$

For $k = 1, 2, \ldots, 28$ we get 28 Milnor spheres.[‡] One of these 28 manifolds is homeomorphic to the sphere S^7.

33.8. Submanifolds. The sphere in $\mathbf{R}^3$ with equation $x^2 + y^2 + z^2 = 1$ is an example of a subset of Euclidean space inheriting the natural structure of a differentiable manifold from $\mathbf{R}^3$, namely the structure of a *submanifold* of $\mathbf{R}^3$. The general definition of a submanifold goes as follows:

Definition. A subset V of a manifold M (Fig. 233) is said to be a *submanifold* if every point $x \in V$ has a neighborhood W in M and a map $\varphi: W \to U$ such that $\varphi(W \cap V)$ is a domain of an affine subspace of the affine space $\mathbf{R}^n$ containing U. The submanifold V itself has the natural structure of a manifold $(W' = W \cap V, \ U' = \varphi(W'))$.

The following fundamental fact is given without proof and will not be used subsequently:

THEOREM. *Every manifold M^n is diffeomorphic to a submanifold of Euclidean space $\mathbf{R}^N$ of sufficiently large dimension (for example $N > 2n$, where $n = \dim M^n$).*

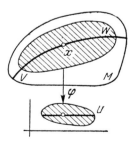

Fig. 233   A submanifold.

[†] A manifold M is said to be *simply connected* if every closed curve in M can be continuously shrunk to a point.

[‡] See E. Brieskorn, *Beispiele zur Differentialtopologie von Singularitäten*, Invent. Math. **2** (1966), 1–14.

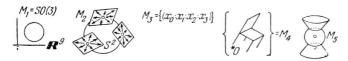

Fig. 234   Examples of three-dimensional manifolds.

Thus the abstract concept of a manifold does not actually comprise a larger class of objects than "k-dimensional surfaces in N-dimensional space." The advantage of the abstract approach is that it includes those cases where no embedding in Euclidean space is specified in advance, and where such a specification would only lead to spurious complications (as in the case of the projective space $\mathbf{R}P^n$). The situation here is the same as for finite-dimensional linear spaces (they are all isomorphic to the coordinate space of points $(x_1, \ldots, x_n)$, but specifying coordinates often merely complicates matters).

33.9. Example. Finally we consider the following five interesting manifolds (Fig. 234):
1) *The group* $M_1 = SO(3)$ *of orthogonal matrices of order* 3 *and determinant* $+1$. Since every matrix of M_1 has 9 elements, M_1 is a subset of the space $\mathbf{R}^9$. It is easy to see that this subset is actually a submanifold.
2) *The set* $M_2 = T_1 S^2$ *of all vectors of length* 1 *tangent to the sphere* S^2 in three-dimensional Euclidean space. As an exercise, the reader should introduce the structure of a differentiable manifold into M_2 (cf. Sec. 34).
3) *The three-dimensional projective space* $M_3 = \mathbf{R}P^3$.
4) *The configuration space* M_4 *of a rigid body* fastened at a fixed point O.
5) *The subset* M_5 *of the space* $\mathbf{R}^6 = {}^{\mathbf{R}}\mathbf{C}^3$ *determined by the equations*
$$z_1^2 + z_2^2 + z_3^2 = 0,$$
$$|z_1|^2 + |z_2|^2 + |z_3|^2 = 2.$$

Problem 1. Which of the manifolds $M_1, \ldots, M_5$ are diffeomorphic?

34. The Tangent Bundle. Vector Fields on a Manifold

With every smooth manifold M there is associated another manifold (of twice the dimension), called the *tangent bundle* of M and denoted by TM.† The whole theory of ordinary differential equations can immediately be carried over to manifolds, with the help of the tangent bundle.

34.1. The tangent space. Given a smooth manifold M, by *the vector* ξ *tangent to* M *at the point* x is meant the equivalence class of curves leaving x, two curves (Fig. 235)
$$\gamma_1 : I \to M, \qquad \gamma_2 : I \to M$$

† A *tangent bundle* is a special case of a *vector bundle*; a still more general concept is that of a *bundle space*. All these notions are basic in topology and analysis, but we confine ourselves here to tangent bundles, which are particularly important in the theory of *ordinary* differential equations.

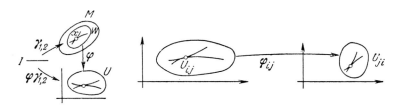

Fig. 235   The tangent vector.

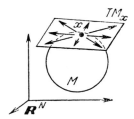

Fig. 236   A tangent space.

being equivalent if their images

$$\varphi\gamma_1 : I \to U, \qquad \varphi\gamma_2 : I \to U$$

on any map are equivalent.

Note that the concept of equivalence of curves does not depend on the choice of the map of the atlas (see Sec. 6): Equivalence on a map φ_i implies equivalence on any other map φ_j, since the transformation φ_{ij} from one map to another is a diffeomorphism.

The set of vectors tangent to M at x has the structure of a linear space, a structure independent of the choice of map (see Sec. 6). This linear space is called the *tangent space to M at x* and is denoted by TM_x. The dimension of TM_x is the same as the dimension of M.

Example 1. Let M^n be a submanifold of the affine space $\mathbf{R}^N$ (Fig. 236). Then TM_x^n can be thought of as an n-dimensional plane in R^N going through x. Here, however, it must be kept in mind that *the tangent spaces to M at distinct points x and y do not intersect*: $TM_x \cap TM_y = \varnothing$.

34.2. The tangent bundle. Consider the union

$$TM = \bigcup_{x \in M} TM_x$$

of the tangent spaces to a manifold M at all points $x \in M$. Then the set TM has the natural structure of a smooth manifold.

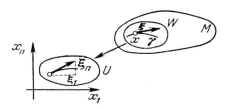

Fig. 237   Coordinates of the tangent vector.

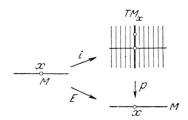

Fig. 238   A tangent bundle.

In fact, consider any map on the manifold M, and let $(x_1, \ldots, x_n) : W \to U \subset \mathbf{R}^n$ (Fig. 237) be local coordinates in a neighborhood W of the point x specifying this map. Every vector $\boldsymbol{\xi}$ tangent to M at a point $x \in W$ is determined by its components $\xi_1, \ldots, \xi_n$ in the indicated coordinate system. In fact, if $\gamma : I \to M$ is a curve leaving x in the direction of $\boldsymbol{\xi}$ at time t_0, then

$$\xi_i = \frac{d}{dt}\bigg|_{t=t_0} x_i(\gamma(t)).$$

Thus every vector $\boldsymbol{\xi}$ tangent to M at a point of the domain W is specified by $2n$ numbers $x_1, \ldots, x_n, \xi_1, \ldots, \xi_n$, the n coordinates of the "point of tangency" x and the n "components" ξ_i. This gives a map of part of the set TM:

$$\psi : TW \to \mathbf{R}^{2n}, \qquad \psi(\boldsymbol{\xi}) = (x_1, \ldots, x_n, \xi_1, \ldots, \xi_n).$$

Different maps of TM corresponding to different maps of the atlas of M are compatible (of class C^{r-1} if M is of class C^r). In fact, let $y_1, \ldots, y_n$ be another local coordinate system on M, and let $\eta_1, \ldots, \eta_n$ be the components of a vector in this system. Then

$$y_i = y_i(x_1, \ldots, x_n), \qquad \eta_i = \sum_{j=1}^{n} \frac{\partial y_i}{\partial x_j} \xi_j \qquad (i = 1, \ldots, n)$$

are smooth functions of x_j and ξ_j. Thus the set TM of all tangent vectors to M acquires a smooth manifold structure of dimension $2n$.

Definition. The manifold TM is called the *tangent bundle* (*space*) of the manifold M.

There exist natural mappings $i : M \to TM$ (the *null section*) and $p : TM \to M$ (*projection*) such that $i(x)$ is the zero vector of TM_x and $p(\boldsymbol{\xi})$ is the point x at which $\boldsymbol{\xi}$ is tangent to M (Fig. 238).

Fig. 239   Parallelizable and nonparallelizable manifolds.

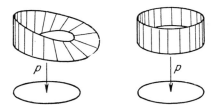

Fig. 240   A bundle which is not a direct product.

Problem 1. Prove that the mappings i and p are differentiable, that i is a diffeomorphism of M onto $i(M)$, and that $p \circ i\colon M \to M$ is the identity mapping.

The preimages of the points $x \in M$ under the mapping $p\colon TM \to M$ are called *fibres* of the bundle TM. Every fibre has the structure of a linear space. The set M is called the *base* of the bundle TM.

34.3. Remarks on parallelizability. The tangent bundle of the affine space $\mathbf{R}^n$ or of a domain $U \subset \mathbf{R}^n$ has the structure of a direct product: $TU = U \times \mathbf{R}^n$. In fact, the tangent vector to U can be specified by a pair (x, ξ), where $x \in U$ and ξ is a vector of the linear space $\mathbf{R}^n$, for which there exists a linear isomorphism with TU_x (Fig. 239). This can be expressed differently by saying that affine space is *parallelizable*, i.e., equality is defined for tangent vectors to the domain $U \subset R^n$ at different points x and y.

The tangent bundle to a manifold M need not be a direct product, and in general we cannot give a reasonable definition of equality for vectors "attached" to different points of M (Fig. 239). The situation here is the same as for a Möbius strip (Fig. 240), which is a tangent bundle with a circle as its base and straight lines as its fibres, but which is not the direct product of the circle and a line.

Definition. A manifold M is said to be *parallelized* if its tangent bundle is expressed as a direct product, i.e., if a diffeomorphism of $TM^n \cong M^n \times \mathbf{R}^n$ carrying TM_x linearly into $x \times \mathbf{R}^n$ is given. A manifold is said to be *parallelizable* if it can be parallelized.

Example 1. Any domain in Euclidean space is naturally parallelized.

Problem 1. Prove that the torus T^n is parallelizable, but not the Möbius strip.

Fig. 241   The hedgehog theorem.

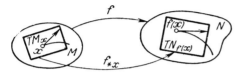

Fig. 242   The derivative of the mapping f at the point x.

*THEOREM. *Only three of the spheres S^n are parallelizable, namely S^1, S^3, and S^7. In particular, the two-dimensional sphere is nonparallelizable*:

$$TS^2 \neq S^2 \times \mathbf{R}^2.$$

This implies, for example, that a hedgehog cannot be combed: At least one quill will be perpendicular to the surface (Fig. 241).

The reader who has solved the problem at the end of Sec. 33.9 will find it easy to prove the nonparallelizability of S^2 (Hint: $\mathbf{RP}^3 \not\cong S^2 \times S^1$). The parallelization of S^1 is obvious, while that of S^3 is an instructive exercise (Hint: S^3 is a group, namely the group of quaternions of modulus 1). A complete proof of the above theorem requires a rather deep penetration into the subject of topology; in fact, the theorem was proved only relatively recently.

Analysts are inclined to regard all bundles as direct products and all manifolds as parallelized. This mistake should be avoided.

34.4. The tangent mapping. Let $f: M \to N$ be a smooth mapping of a manifold M into a manifold N (Fig. 242), and let f_{*x} denote the induced mapping of the tangent spaces. The mapping f_{*x} $(= f_*|_x)$ is defined as in Sec. 6.3, and is a linear mapping of one linear space into another:

$$f_{*x}: TM_x \to TN_{f(x)}. \tag{1}$$

Let x vary over M. Then (1) defines a mapping

$$f_*: TM \to TN, \qquad f_*|_{TM_x} = f_{*x}$$

of the tangent bundle of M into the tangent bundle of N. This mapping is

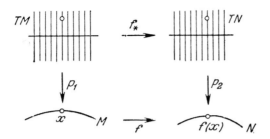

Fig. 243   The tangent mapping.

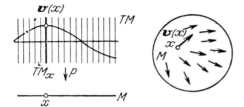

Fig. 244   A vector field.

differentiable (why?) and maps the fibres of TM linearly into the fibres of TN (Fig. 243).

The mapping f_{*x} is called the *tangent mapping* of f (the notation $Tf\colon TM \to TN$ is also used).

Problem 1. Let $f\colon M \to N$ and $g\colon N \to K$ be smooth mappings, with composition $g \circ f\colon M \to K$. Prove that $(g \circ f)_* = g_* \circ f_*$, i.e., that

$$
\begin{array}{ccc}
 & N & \\
f \nearrow & & \searrow g \\
M \xrightarrow[g \circ f]{} & & K
\end{array}
\quad \Rightarrow \quad
\begin{array}{ccc}
 & TN & \\
f_* \nearrow & & \searrow g_* \\
TM \xrightarrow[(g \circ f)_*]{} & & TK.
\end{array}
$$

Comment on terminology. In analysis this formula is called the rule for differentiation of a composite function, while in algebra it is called the *(covariant) functoriality* of the transition to the tangent mapping.

34.5. Vector fields. Let M be a smooth manifold (of class C^{r+1}), with tangent bundle TM (Fig. 244).

Definition. By a *vector field*† $\mathbf{v}$ (of class C^r) on M is meant a smooth mapping $\mathbf{v}\colon M \to TM$ (of class C^r) such that the mapping $p \circ \mathbf{v}\colon M \to M$ is the

---

† The term *section of the tangent bundle* is also used.

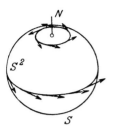

Fig. 245   A velocity field.

identity mapping E or equivalently such that the diagram

is commutative, i.e., $p(\mathbf{v}(x)) = x$.

Remark. If M is a domain of the space $\mathbf{R}^n$ with coordinates $x_1, \ldots, x_n$, this definition coincides with the old one (Sec. 1.4). However, the present definition involves no special system of coordinates.

Example. Consider the family of rotations g^t of the sphere S^2 through the angle t about the axis SN (Fig. 245). Every point of the sphere $x \in S^2$ describes a curve (a parallel of latitude) under the rotation, with velocity

$$\mathbf{v}(x) = \frac{d}{dt}\Big|_{t=0} g^t x \in TS_x^2.$$

This gives the mapping $\mathbf{v}: S^2 \to TS^2$, where obviously $p \circ \mathbf{v} = E$, i.e., $\mathbf{v}$ is a vector field on S^2.

In general, a one-parameter group of diffeomorphisms $g^t: M \to M$ of the manifold M gives rise to a vector field of the phase velocity on M, in precisely the same way as in Sec. 1.4. The whole local theory of (nonlinear) ordinary differential equations can immediately be carried over to manifolds, since we were careful at the time (in Sec. 6) to keep our basic concepts independent of the coordinate system. In particular, the basic local theorem on rectifiability of a vector field and the local theorems on existence, uniqueness, continuity, and differentiability with respect to initial conditions carry over to manifolds. The specific character of the manifold comes to the fore only in considering nonlocal problems. The simplest of these problems concerns the existence of solutions or the existence of a phase flow with a given phase velocity field.

35. The Phase Flow Determined by a Vector Field

The theorem to be proved below is the simplest theorem of the qualitative theory of differential equations, giving conditions under which it makes sense to ask about the behavior of solutions of a differential equation on an infinite time interval. In particular, the theorem implies the continuity and differentiability of the solution with respect to the initial data in the large (i.e., on any finite time interval). The theorem is also useful as a model of the technique of constructing diffeomorphisms. For example, we can use the theorem to prove that every closed manifold having a smooth function with only two critical points is homeomorphic to a sphere.

35.1. Theorem. *Let M be a smooth manifold (of class C^r, $r \geqslant 2$), and let* $\mathbf{v}: M \to TM$ *be a vector field (Fig. 246). Moreover, let the vector $\mathbf{v}(x)$ be different from the zero vector of TM_x only in a compact subset K of the manifold M. Then there exists a one-parameter group of diffeomorphisms $g^t: M \to M$ for which $\mathbf{v}$ is the phase velocity field:*

$$\frac{d}{dt}g^t x = \mathbf{v}(g^t x). \tag{1}$$

COROLLARY 1. *Every vector field $\mathbf{v}$ on a compact manifold M is the phase velocity field of a one-parameter group of diffeomorphisms.*

In particular, under the conditions of the theorem or those of Corollary 1, we have

COROLLARY 2. *Every solution of the differential equation*

$$\dot{x} = \mathbf{v}(x), \qquad x \in M \tag{2}$$

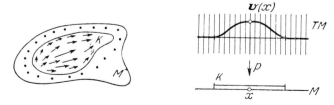

Fig. 246   A vector field vanishing outside a compact set K.

can be extended indefinitely forward and backward, with the value of the solution $g^t x$ at time t depending smoothly on t and the initial condition x.

Remark. The compactness condition cannot be dropped.

Example 1. If $M = \mathbf{R}$, $\dot{x} = x^2$ (see Sec. 3.5), the solutions cannot be extended indefinitely.

Example 2. $M = \{x : 0 < x < 1\}$, $\dot{x} = 1$.

We now proceed to prove the theorem.

35.2. Construction of the diffeomorphisms g^t for small t. *For every point $x \in M$ there exists an open neighborhood $U \subset M$ and a number $\varepsilon > 0$ such that given any point $y \in U$ and any t with $|t| < \varepsilon$, the solution $g^t y$ of equation (2) satisfying the initial condition y (at $t = 0$) exists, is unique, depends differentiably on t and y, and satisfies the condition*

$$g^{t+s} y = g^t g^s y$$

if $|t| < \varepsilon, |s| < \varepsilon, |t + s| < \varepsilon$.

In fact, the point x has an image on some map, and our assertion has been proved for equations in a domain of affine space (see Chaps. 2 and 4).†

Thus the compact set K is covered by neighborhoods U from which we can select a finite covering $\{U_i\}$. Let ε_i be the corresponding numbers ε, and choose $\varepsilon_0 = \min \varepsilon_i > 0$. Then for $|t| < \varepsilon_0$ we can define diffeomorphisms $g^t : M \to M$ *in the large* such that $g^t x = x$ for x outside K and $g^{t+s} = g^t g^s$ if $|t|, |s|, |t + s| < \varepsilon_0$. In fact, although the solutions of equation (2) with the initial condition x (for $t = 0$) defined by using different maps are different *a priori*, they coincide for $|t| < \varepsilon_0$ because of the choice of ε_0 and the local uniqueness theorem. Moreover, by the local theorem on differentiability, the point $g^t x$ depends differentiably on t and x, and since $g^t g^{-t} = E$, the mapping $g^t : M \to M$ is a diffeomorphism. Note also that

$$\frac{d}{dt}\bigg|_{t=0} g^t x = \mathbf{v}(x). \tag{3}$$

35.3. Construction of g^t for arbitrary t. Let t be represented in the form $(n\varepsilon_0/2) + r$, where n is an integer and $0 \leqslant r < \varepsilon_0/2$ (this representation

---

† The proof of the uniqueness requires a slight additional argument: It must be verified that uniqueness of the solution with given initial conditions on every fixed map implies uniqueness on the manifold. Uniqueness may well fail on a nonseparable manifold (consider, for example, the equation $\dot{x} = 1, \dot{y} = 1$ on the manifold obtained from the lines $\{x\}$ and $\{y\}$ by identifying points with equal negative coordinates). However, if the manifold M is separable, then the uniqueness proof of Sec. 7.7 goes through. (The separability is used to prove the coincidence of the values of the solutions $\varphi_1(T)$ and $\varphi_2(T)$ at the first point T after which they no longer coincide.)

exists and is unique). The diffeomorphisms $g^{\varepsilon_0/2}$ and g^r have already been defined. Writing $g^t = (g^{\varepsilon_0/2})^n g^r$, we get a diffeomorphism of M onto M. For $|t| < \varepsilon_0/2$ the new definition agrees with that of Sec. 35.2, and hence (3) holds. Moreover, it is easy to see that

$$g^{s+t} = g^s g^t \tag{4}$$

for arbitrary s and t.

In fact, let

$$s = m\frac{\varepsilon_0}{2} + p, \qquad t = n\frac{\varepsilon_0}{2} + q, \qquad s + t = k\frac{\varepsilon_0}{2} + r.$$

Then the left and right-hand sides of (4) become $(g^{\varepsilon_0/2})^k g^r$ and $(g^{\varepsilon_0/2})^m g^p (g^{\varepsilon_0/2})^n g^q$. Two cases are possible:

1) $m + n = k, \qquad p + q = r;$

2) $m + n = k - 1, \qquad p + q = r + \dfrac{\varepsilon_0}{2}.$

But the diffeomorphisms $g^{\varepsilon_0/2}$, g^p, and g^q commute, since $|p| < \varepsilon_0/2$, $|q| < \varepsilon_0/2$. This implies (4) in both the first and second cases $(g^{\varepsilon_0/2} g^r = g^p g^q$ since $|p|, |q|, |r| < \dfrac{\varepsilon_0}{2}$, $p + q = \dfrac{\varepsilon_0}{2} + r).$

We must still verify that the point $g^t x$ depends differentiably on t and x. This follows, for example, from the fact that $g^t = (g^{t/N})^N$, while $g^{t/N} x$ depends differentiably on t and x for sufficiently large N, by Sec. 35.2.

Thus $\{g^t\}$ is a one-parameter group of diffeomorphisms of the manifold M, and $\mathbf{v}$ is the corresponding field of the phase velocity. The proof of Theorem 35.1 is now complete. ∎

35.4. Remark. It is a simple consequence of Theorem 35.1 that *every solution of the nonautonomous equation*

$$\dot{x} = \mathbf{v}(x, t), \qquad x \in M, \quad t \in \mathbf{R}$$

defined by a time-dependent vector field $\mathbf{v}$ on a compact manifold M can be extended indefinitely.

In particular, this explains why we can extend solutions of the linear equation

$$\dot{\mathbf{x}} = \mathbf{v}(\mathbf{x}, t), \qquad \mathbf{v}(\mathbf{x}, t) = A(t)\mathbf{x}, \quad t \in \mathbf{R}, \quad \mathbf{x} \in \mathbf{R}^n \tag{5}$$

indefinitely. In fact, we will regard $\mathbf{R}^n$ as the affine part of the projective space $\mathbf{RP}^n$, where the latter is obtained from its affine part by adjoining the plane at infinity: $\mathbf{RP}^n = \mathbf{R}^n \cup \mathbf{RP}^{n-1}$. Let $\mathbf{v}$ be a linear vector field in $\mathbf{R}^n$, so that $\mathbf{v}(\mathbf{x}) = A\mathbf{x}$. Then we can easily prove the following

Fig. 247   Extension of a linear vector field onto projective space.

Fig. 248   Behavior of the extension of the field near the plane at infinity.

LEMMA. *The vector field* **v** *on* $\mathbf{R}^n$ *can be uniquely extended to a smooth field* **v**′ *on* $\mathbf{RP}^n$. *The field* **v**′ *on the plane at infinity* $\mathbf{RP}^{n-1}$ *is tangent to* $\mathbf{RP}^{n-1}$.

In particular, suppose that (for every t) we extend the field $\mathbf{v}(t)$ specifying (5) to a field $\mathbf{v}'(t)$ on $\mathbf{RP}^n$. Consider the equation

$$\dot{\mathbf{x}} = \mathbf{v}'(\mathbf{x}, t), \qquad \mathbf{x} \in \mathbf{RP}^n, \quad t \in \mathbf{R}. \tag{6}$$

Since projective space is compact, every solution of (6) can be extended indefinitely (Fig. 247). A solution initially belonging to $\mathbf{RP}^{n-1}$ always stays in $\mathbf{RP}^{n-1}$, since the field $\mathbf{v}'$ is tangent to $\mathbf{RP}^{n-1}$. By the uniqueness theorem, the solutions of the equation with initial conditions in $\mathbf{R}^n$ remain in $\mathbf{R}^n$ for all t. But equation (6) is of the form (5) in $\mathbf{R}^n$. Thus every solution of (5) can be extended indefinitely.

Problem. Prove the lemma.

Solution 1. Let $x_1, \ldots, x_n$ be affine coordinates in $\mathbf{RP}^n$ and $y_1, \ldots, y_n$ other affine coordinates such that

$$y_1 = x_1^{-1}, \qquad y_k = x_k x_1^{-1}, \qquad k = 2, \ldots, n.$$

Then the equation of $\mathbf{RP}^{n-1}$ is just $y_1 = 0$ in the new coordinates. The differential equation (5)

$$\frac{dx_i}{dt} = \sum_{j=1}^{n} a_{ij} x_j, \qquad i = 1, \ldots, n$$

takes the form

$$\frac{dy_1}{dt} = -y_1\left(a_{11} + \sum_{k>1} a_{1k} y_k\right),$$

$$\frac{dy_k}{dt} = a_{k1} + \sum_{l>1} a_{kl} y_l - y_k\left(a_{11} + \sum_{l>1} a_{1l} y_l\right), \qquad k > 1$$

in the new coordinates (Fig. 248). From these formulas, valid for $y_1 \neq 0$, it is clear how to complete the definition of the field at $y_1 = 0$. For $y_1 = 0$ we get $dy_1/dt = 0$, thereby proving the lemma.

Solution 2. An affine transformation can be regarded as a projective transformation, leaving the plane at infinity (but not its points) fixed. In particular, the linear transformations e^{At} can be extended to diffeomorphisms of projective space leaving the plane at infinity fixed. These diffeomorphisms form a one-parameter group, with $\mathbf{v}'$ as its phase velocity field.

36. The Index of a Singular Point of a Vector Field

We now consider a few simple applications of topology to the study of differential equations.

36.1. The index of a curve. We begin with some intuitive considerations which will be backed up later by exact definitions and proofs (see Sec. 36.6).

Consider a vector field specified in an oriented Euclidean plane. Suppose we are given an oriented closed curve in the plane, which does not go through any singular points of the field (Fig. 249), and suppose a point makes one circuit around the curve in the positive direction. Then the field vector at the point in question will rotate continuously as the point moves around the curve.[†] When the point returns to its original position, having

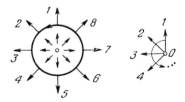

Fig. 249   A curve of index 1.

[†] To keep track of the revolutions of the vector, it is convenient to refer all vectors to a single point O, following the natural parallelization of the plane.

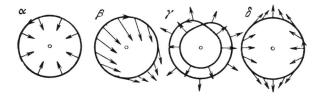

Fig. 250   Curves with various indices.

gone around the curve, the vector also returns to its original position, but in doing so, it may make several revolutions in one direction or the other. The number of revolutions made by the field vector in traversing the curve once is called the *index* of the curve. Here the number of revolutions is taken with the plus sign if the vector rotates in the direction specified by the orientation of the plane (from the first basis vector to the second), and with the minus sign otherwise.

Example 1. The indices of the curves α, β, γ, and δ in Fig. 250 are $1, 0, 2$, and -1 respectively.

Example 2. Let O be a nonsingular point of the field. Then the index of every curve lying in a sufficiently small neighborhood of O equals zero. In fact, the direction of the field at O is continuous and hence changes by less than $\pi/2$, say, in a sufficiently small neighborhood of O.

Problem 1. Suppose we specify a vector field in the plane $\mathbf{R}^2 = {}^{\mathbf{R}}\mathbf{C}$ without the point O by the formula $\mathbf{v}(z) = z^n$, where n is an integer which is not necessarily positive. Calculate the index of the circle $z = e^{i\varphi}$, oriented in the direction of increasing φ (the plane is oriented by the frame $1, i$).

Ans. n.

36.2. Properties of the index and their implications.

PROPERTY 1. *The index of a closed curve does not change under continuous deformation, as long as the curve does not go through any singular points.*

In fact, the direction of the field vector changes continuously away from the singular points. Therefore the number of revolutions also depends continuously on the curve, and hence must be constant, being an integer. ∎

PROPERTY 2. *The index of a curve does not change under continuous deformation of the vector field, provided only that there are no singular points of the field on the curve during the whole course of the deformation.*

These two properties, which are quite obvious intuitively,[†] have a number of deep implications:

THEOREM 1. *Given a vector field in the plane, let D be a circular disk and S its boundary.* [‡] *If the index of the curve S is nonzero, then there is at least one singular point inside D.*

Proof. If there are no singular points in D, then S can be deformed continuously inside D without going through any singular points, so that after the deformation we get a curve arbitrarily close to a point O in D (we can even deform S into the point O). The index of the resulting small curve equals zero. But the index does not change under deformation, and hence it must originally have been equal to zero, contrary to hypothesis. ∎

Problem 1. Prove that the system of differential equations
$$\dot{x} = x + P(x, y), \qquad \dot{y} = y + Q(x, y),$$
where P and Q are functions bounded in the whole plane, has at least one equilibrium position.

THEOREM 2 (**Fundamental theorem of algebra**). *Every equation*
$$z^n + a_1 z^{n-1} + \cdots + a_n = 0 \tag{1}$$
has at least one complex root.

First we prove the following

LEMMA. *Let* $\mathbf{v}$ *be the vector field in the plane of the complex variable z given by the formula*
$$\mathbf{v}(z) = z^n + a_1 z^{n-1} + \cdots + a_n,$$
so that the singular points of $\mathbf{v}$ *are just the roots of equation* (1). *Then the index in the field* $\mathbf{v}$ *of a circle of sufficiently large radius equals n.* [§]

Proof. In fact, the formula
$$\mathbf{v}_t(z) = z^n + t(a_1 z^{n-1} + \cdots + a_n), \qquad 0 \leqslant t \leqslant 1$$
defines a continuous deformation of the original field into the field z^n. If $r > 1 + |a_1| + \cdots + |a_n|$, then $r^n > |a_1| r^{n-1} + \cdots + |a_n|$. Hence there are no singular points on a circle of radius r during the whole course of the deformation. It follows from Property 2 that the index of this circle is the

---

[†] The accurate formulation and proof of these assertions requires some topological technique, namely the use of homotopies, homologies, or something similar (to this end, we will use Green's formula below). See e.g., W. G. Chinn and N. E. Steenrod, *First Concepts of Topology*, New York (1966).
[‡] We can also consider the more general case where D is any plane domain bounded by a simple closed curve S.
[§] Here we use the same orientation as in Sec. 36.1, Problem 1.

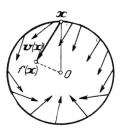

Fig. 251   Mapping of a disk into itself.

same in the original field as in the field z^n. But the index equals n in the field z^n. ∎

Proof of Theorem 2. Let r be the same as in the proof of the lemma. Then, by Theorem 1 and the lemma, there is at least one singular point of the vector field, i.e., at least one root of equation (1), inside the disk of radius r. ∎

THEOREM 3 (**Fixed point theorem**). *Every smooth† mapping* $f: D \to D$ *of a closed disk into itself has at least one fixed point.*

Proof. We take the plane of the disk D to be a linear space having its origin at the center of the disk (Fig. 251). The fixed points of the mapping f are just the singular points of the vector field $\mathbf{v}(\mathbf{x}) = f(\mathbf{x}) - \mathbf{x}$. If there are no singular points in D, then there are none on the circle bounding D. This circle has index 1 in the field $\mathbf{v}$. In fact, there exists a continuous deformation of the field $\mathbf{v}$ into the field $-\mathbf{x}$ such that there are no singular points on the circle during the whole course of the deformation (for example, we need only set $\mathbf{v}_t(\mathbf{x}) = tf(\mathbf{x}) - \mathbf{x}, 0 \leqslant t \leqslant 1$). Hence the circle has the same index in both fields $\mathbf{v}_0 = -\mathbf{x}$ and $\mathbf{v}_1 = \mathbf{v}$. But a simple direct calculation shows that the index of the circle $|\mathbf{x}| = r$ in the field $-\mathbf{x}$ equals 1. To complete the proof, we again use Theorem 1 to deduce that there is at least one singular point of the field $\mathbf{v}$, i.e., at least one fixed point of the mapping f, inside the disk. ∎

36.3. The index of a singular point. Let O be an isolated singular point of a vector field in the plane, i.e., suppose there are no other singular points in some neighborhood of O. Consider a circle of sufficiently small radius centered at O. Suppose the plane is oriented and let the orientation of the circle be positive (as in Sec. 36.1).

THEOREM. *The index of a circle of sufficiently small radius centered at an isolated*

---

† The theorem is valid for any continuous mapping, but here we prove the theorem only under the assumption of smoothness (see Sec. 36.6).

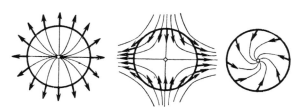

Fig. 252 The indices of simple singular points equal ±1.

singular point O does not depend on the radius of the circle, provided only that the radius is sufficiently small.

Proof. Any two such circles can be continuously deformed into each other without going through singular points. ∎

Note also that instead of a circle, we can choose any other curve going around O once in the positive direction.

Definition. The index of any (and hence every) sufficiently small positively oriented circle centered at an isolated singular point of a vector field is called the *index of the singular point.*

Examples. Suppose the singular point is a node, saddle point, or focus (or center). Then the index of the singular point is +1, −1, or +1 repectively (Fig. 252).

A singular point of a vector field is said to be *simple* if the operator of the linear part of the field at the point is nondegenerate. The class of singular points in the plane consists of nodes, saddle points, foci, and centers. Thus the index of such a singular point is always ±1.

Problem 1. Construct a vector field with a singular point of index n.

Hint. See, for example, the problem in Sec. 36.1.

Problem 2. Prove that the index of a singular point is independent of the choice of orientation of the plane.

Hint. Changing the orientation simultaneously changes both the positive direction of traversing the circle and the positive direction of counting the number of revolutions.

36.4. Index of a curve in terms of indices of singular points. Let D be a compact domain bounded by a simple curve S in the oriented plane. Suppose S has the standard orientation of the boundary of D, i.e., suppose D lies to the left of an observer traversing S in the positive direction. This means that the positive orientation of the plane is given by the dihedral made up of the velocity vector along S and the normal vector directed inside D.

Now suppose we are given a vector field in the plane, with no singular points on the curve S and only a finite number of singular points inside the domain D.

THEOREM. *The index of the curve S equals the sum of the indices of the singular points of the field lying inside D.*

First we prove that the index of a curve has the following additivity property:

LEMMA. *Given two oriented curves γ_1 and γ_2 going through the same point, let $\gamma_1 + \gamma_2$ be the new oriented curve obtained by traversing first γ_1 and afterwards γ_2. Then the index of $\gamma_1 + \gamma_2$ equals the sum of the indices of γ_1 and γ_2.*

Proof. The field vector makes n_1 turns in going around γ_1 and n_2 more turns in going around γ_2, and hence $n_1 + n_2$ turns in all.   ∎

Proof of the theorem. We partition D into parts D_i such that there is no more than one singular point of the field inside each part (Fig. 253), and no singular points at all on the boundaries of the parts. Moreover, we assign each of the curves γ_i bounding the parts D_i the orientation appropriate to the boundary (Fig. 253). Then, by the lemma,

$$\text{ind} \sum_i \gamma_i = \text{ind} \, S + \sum_j \text{ind} \, \delta_j,$$

where the closed curve δ_j is made up of a part of the boundary of D_j lying inside D and is traversed twice in opposite directions. The index of each curve δ_j equals 0, since δ_j can be contracted into a point without passing through singular points (see Sec. 36.2). The index of the curve γ_i equals the index of the singular point surrounded by γ_i (or 0 if the domain D_i surrounded by γ_i contains no singular points).   ∎

Problem 1. Let $p(z)$ be a polynomial of degree n in a complex variable z, and let D be a domain in the z-plane bounded by a curve S. Suppose there are no zeros of the polynomial on S. Prove that the number of zeros of the polynomial inside D (with multiplicities taken into account) equals the index of the curve S in the field $\mathbf{v} = p(z)$, i.e., the

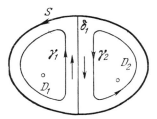

Fig. 253   The index of the curve S equals the sum of the indices of the curves γ_1 and γ_2.

number of revolutions (winding number) of the curve $p(S)$ around the origin.

Comment. This gives a way of solving the Routh-Hurwitz problem of Sec. 23.4: *Find the number n_- of zeros of a given polynomial in the left half-plane.* To this end, we consider a half-disk of sufficiently large radius in the left half-plane with its center at the point $z = 0$ and its diameter along the imaginary axis. The number of zeros in the left half-plane equals the index of the boundary S of the half-disk (if the radius is large enough and if the polynomial has no purely imaginary zeros). To calculate the index of the curve S, we need only find the number of revolutions ν around the origin of the image of the imaginary axis (oriented from $-i$ to $+i$). In fact, it is easily verified that

$$n_- = \text{ind } S = \nu + \frac{n}{2},$$

since the image under the mapping p of a semicircle of sufficiently large radius makes approximately $n/2$ revolutions around the origin (a number closer to $n/2$, the larger the radius).

In particular, all the zeros of a polynomial of degree n lie in the left half-plane if and only if the point $p(it)$ goes around the origin $n/2$ times (in the direction from 1 to i) as t varies from $-\infty$ to $+\infty$.

36.5. The sum of the indices of singular points on a sphere.

Problem 1. Prove that the index of a singular point of a vector field in the plane is invariant under a diffeomorphism.

Thus the index is a geometric concept which is independent of the coordinate system. This fact allows us to define the index of a singular point not only in the plane but also on any two-dimensional manifold. In fact, we need only consider the index of the singular point on any map, and the index will then be the same on the other maps.

Example 1. Consider the sphere $x^2 + y^2 + z^2 = 1$ in Euclidean three-dimensional space. The vector field of the velocity of rotation about the z-axis ($\dot{x} = y$, $\dot{y} = -x$, $\dot{z} = 0$) has two singular points, at the north and south poles (Fig. 254), each of index $+1$.

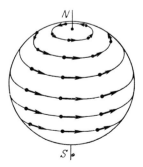

Fig. 254   A vector field on the sphere with two singular points of index 1.

Suppose we are given a vector field on the sphere with only isolated singular points. Then there are only a finite number of such points, since the sphere is compact.

*THEOREM. *The sum of the indices of all the singular points of a field on the sphere is independent of the choice of the field.*

It is clear from the above example that *this sum equals* 2.

Idea of the proof. Consider a map of the sphere covering the whole sphere except for one point, which we call the pole. Then consider the field of the basis vector $\mathbf{e}_1$ in the Euclidean plane of this map, and carry the field over to the sphere. This gives a field on the sphere (defined except at the pole) which we continue to denote by $\mathbf{e}_1$.

Now consider the map of a neighborhood of the pole. In the plane of this map we can also draw the vector field $\mathbf{e}_1$ on the sphere, defined except at one point O. The appearance of this field is shown in Fig. 255.

LEMMA. *The index of a closed curve going once around the point O in the planar field just constructed equals* 2.

Proof. We need only carry out explicitly the operations described above, choosing for the two maps, for example, maps of the sphere under stereographic projection (Fig. 222). Parallel lines on one map then go into the circles shown in Fig. 255 on the second map, from which it is clear that the index equals 2.   ∎

Completion of the proof. Consider a vector field $\mathbf{v}$ on the sphere, choosing a nonsingular point of the field as the pole. Then all the singular points of the field have images on the map of the complement of the pole. The sum of the indices of all the singular points of the field equals the index of a circle of sufficiently large radius in the plane of this map (by Theorem 36.4). We now carry this circle over to the sphere, and then from the sphere to the map of a neighborhood of the pole. The resulting circle on the latter map has index 0 in the field under consideration, since the pole is a nonsingular point of the

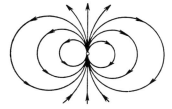

Fig. 255   The vector field $\mathbf{e}_1$, parallel on one map of the sphere, but drawn on another map.

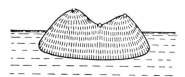

Fig. 256   On every island the sum of the number of peaks and the number of valleys is 1 greater than the number of passes.

field. Staying on the new map, we can interpret the index of a circle on the first map as the "number of revolutions of the field **v** *relative to the field* **e**$_1$" in going once around the circle. This number equals $+2$, since as we go around the circle surrounding the point O on the new map in the positive direction for the first map, the image field **e**$_1$ on the new map makes -2 revolutions while the field **v** makes 0 revolutions. ∎

*Problem 2. Let $f: S^2 \to \mathbf{R}^1$ be a smooth function on the sphere, all of whose critical points are simple (i.e., whose second differential is nondegenerate at every critical point). Prove that

$$m_0 - m_1 + m_2 = 2,$$

where m_i is the number of critical points whose Hessian matrix $(\partial^2 f/\partial x_i \partial x_j)$ has i negative eigenvalues. In other words, *the number of minima minus the number of saddle points plus the number of maxima always equals 2.*

For example, the total number of mountain peaks on earth plus the total number of valleys is 2 greater than the number of passes. If we restrict ourselves to an island or a continent, i.e., if we consider functions on a disk with no singular points on its boundary, then $m_0 - m_1 + m_2 = 1$ (Fig. 256).

Hint. Consider the gradient of the function f.

*Problem 3. Prove *Euler's theorem on polyhedra*, which asserts that

$$\alpha_0 - \alpha_1 + \alpha_2 = 2$$

for every bounded convex polyhedron with α_0 vertices, α_1 edges, and α_2 faces.

Hint. This problem can be reduced to the preceding problem.

*Problem 4. Prove that the sum χ of the indices of the singular points of a vector field on any compact two-dimensional manifold is independent of the field.

The number χ in question is called the *Euler characteristic* of the manifold. For example, we have just seen that the Euler characteristic $\chi(S^2)$ of the sphere equals 2.

Problem 5. Find the Euler characteristic of the torus, of the pretzel, and of the sphere with n handles (Fig. 232).

Ans. $0, -2, 2 - 2n$.

*Problem 6. Extend the results of Problems 2 and 3 from the sphere to any compact two-dimensional manifold, i.e., prove that

$$m_0 - m_1 + m_2 = \alpha_0 - \alpha_1 + \alpha_2 = \chi(M).$$

36.6. A more rigorous approach. We now give an exact definition of the *number of revolutions* or *winding number* of a vector field. Let $\mathbf{v}$ be a smooth vector field defined in a domain U of the plane (x_1, x_2), with components $v_1(x_1, x_2)$ and $v_2(x_1, x_2)$, where the system of coordinates x_1, x_2 specifies an orientation and a Euclidean structure in the plane. Let U' denote the domain obtained from U by deleting the singular points of the field, and let

$$f: U' \to S^1, \qquad f(x) = \frac{\mathbf{v}(x)}{|\mathbf{v}(x)|}$$

be a mapping of U' onto a circle. This mapping is smooth (since singular points of the field have been excluded). Given any point $x \in U'$, we can introduce an angular coordinate φ on a circle in a neighborhood of the image $f(x)$ of the point x. This gives a smooth real function $\varphi(x_1, x_2)$ defined in a neighborhood of x. Calculating the total differential of φ, we get

$$d\varphi = d \arctan \frac{v_2}{v_1} = \frac{v_2 dv_1 - v_1 dv_2}{v_1^2 + v_2^2} \tag{2}$$

for $v_1 \neq 0$. The left and right-hand sides of (2) are also equal for $v_1 = 0$, $v_2 \neq 0$. Thus although the function φ is defined only locally and only to within an integral multiple of 2π, the differential of φ is a well-defined smooth differential form in the whole domain U'. We denote this form by $d\varphi$.

Definition. By the *index of an oriented closed curve* $\gamma: S^1 \to U'$ we mean the integral of the form (2) along γ divided by 2π:

$$\operatorname{ind} \gamma = \frac{1}{2\pi} \oint_\gamma d\varphi \tag{3}$$

We can now give rigorous proofs of the various theorems appearing above. For example, the proof of Theorem 36.4 goes as follows:

Proof. Let D be the domain with boundary S inside which the given field $\mathbf{v}$ has only a finite number of singular points, and let D' be the domain obtained from D by deleting small circular neighborhoods of the singular points. Then the boundary of D', with orientation taken into account, is just

$$\partial D' = S - \sum_i S_i,$$

where S_i is a circle going around the ith singular point in the positive direction (Fig. 257). Applying Green's formula to the domain D' and the

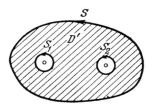

Fig. 257   The domain to which Green's formula is applied.

integral (3), we get

$$\iint_{D'} 0 = \oint_S d\varphi - \sum_i \oint_{S_i} d\varphi.$$

The left-hand side vanishes, since the form (2) is locally a total differential. But then ind $S = \sum$ ind S_i, because of the definition (3).   ∎

*Problem 1. Prove that the index of a closed curve is an integer.

*Problem 2. Give complete proofs of the assertions in Secs. 36.1–36.3.

36.7. The multidimensional case.
The multidimensional generalization of the concept of the *winding number* is the *degree of a mapping*, by which is meant the number of preimages counted with due regard for the signs determined by the orientations. For example, the degree of the mapping of an oriented circle onto another oriented circle shown in Fig. 258 equals 2, since the number of preimages of the point y, with sign taken into account, equals $1 + 1 - 1 + 1 = 2$.

To give a general definition, we proceed as follows. Let $f: M_1^n \to M_2^n$ be a smooth mapping of one n-dimensional oriented manifold onto another such manifold. A point $x \in M_1^n$ in the preimage manifold is called a *regular point* if the derivative of the mapping f at the point x is a nonsingular linear operator $f_{*x}: TM_{1x}^n \to TM_{2f(x)}^n$. For example, the point x in Fig. 258 is regular, but not the point x'.

Definition. By the *degree of the mapping f at a regular point x* is meant the number $\deg_x f$ equal to $+1$ or -1 depending on whether f_{*x} carries the given orientation of the space TM_{1x}^n into the given orientation of the space TM_{2x}^n or into the opposite orientation.

Problem 1. Prove that the degree of a linear automorphism $A: \mathbf{R}^n \to \mathbf{R}^n$ is the same at all points and equals

$$\deg_x A = \operatorname{sgn} \det A = (-1)^{m_-},$$

where m_- is the number of eigenvalues of the operator A with a negative real part.

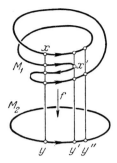

Fig. 258   A mapping of degree 2.

Problem 2. Given a linear automorphism $A: \mathbf{R}^n \to \mathbf{R}^n$ in Euclidean space, define a mapping of the unit sphere onto itself by the formula $f(x) = A(x)/|Ax|$. Find the degree of the mapping f at the point x.

Ans. $\deg_x f = \deg A$.

Problem 3. Let $f: S^{n-1} \to S^{n-1}$ be a mapping carrying every point of the sphere into the diametrically opposite point. What is the degree of f at the point x?

Ans. $\deg_x f = (-1)^n$.

Problem 4. Let $A: \mathbf{C}^n \to \mathbf{C}^n$ be a $\mathbf{C}$-linear automorphism. Find the degree of its decomplexification $^{\mathbf{R}}A$.

Ans. $+1$.

Now consider any point y of the image manifold M_2^n. The point $y \in M_2^n$ is said to be a *regular value of the mapping f* if all the points of its complete preimage $f^{-1}y$ are regular. For example, the point y in Fig. 258 is a regular value, but not the point y'.

THEOREM. *If the manifolds M_1^n and M_2^n are compact and connected, then*
1) *Regular values exist;*
2) *The number of points in the preimage of a regular value is finite;*
3) *The sum of the degrees of the mapping at all the points of the preimage of a regular value does not depend on the particular regular value under consideration.*

The proof of this theorem is quite complicated, and can be found in the literature on topology. †

Remark 1. Actually almost all points of the manifold M_2^n are regular values, i.e., the nonregular values form a set of measure zero.

Remark 2. The compactness condition is essential not only for the second assertion of the theorem, but also for the third assertion. (For example,

† See H. I. Levine, *Singularities of Differentiable Mappings*, Math. Inst. Univ. Bonn (1959), Sec. 6.3.

consider the embedding of the negative real axis in the full real axis.)

Remark 3. The number of points in the preimage (without regard for sign) can be different for different regular values (for example, in Fig. 258 the value y has four such points, while y'' has precisely two).

Definition. By the *degree of the mapping* f is meant the sum of the degrees of f at all the points of the preimage of a regular value of f:

$$\deg f = \sum_{x \in f^{-1}y} \deg_x f.$$

Problem 5. Find the degree of the mapping of the circle $|z| = 1$ onto itself given by the formula $f(z) = z^n,\ n = 0,\ \pm 1,\ \pm 2,\ \ldots$

Ans. n.

Problem 6. Find the degree of the mapping of the unit sphere in Euclidean space $\mathbf{R}^n$ onto itself given by the formula $f(x) = Ax/|Ax|$, where $A: \mathbf{R}^n \to \mathbf{R}^n$ is a nonsingular linear operator.

Ans. $\deg f = \operatorname{sgn} \det A$.

Problem 7. Find the degree of the mapping of the complex projective line $\mathbf{CP}^1$ onto itself given by the formula
a) $f(z) = z^n$;   b) $f(z) = z^{-n}$.

Ans. a) $|n|$;   b) $-|n|$.

Problem 8. Find the degree of the mapping of the complex line $\mathbf{CP}^1$ onto itself given by a polynomial of degree n.

*$*Problem 9.*$ Let $f: U' \to S^1$ be the mapping constructed in Sec. 36.6 with the help of a vector field $\mathbf{v}$ in a domain U', let $\gamma: S^1 \to U'$ be a closed curve, and let $h = f \circ \gamma: S^1 \to S^1$. Prove that the index of γ as defined in Sec. 36.6 coincides with the degree of h:

ind $\gamma = \deg h$.

Definition. By the *index of an isolated singular point* O of a vector field $\mathbf{v}$ defined in a domain of Euclidean space $\mathbf{R}^n$ containing O is meant the degree of the mapping h corresponding to the field, i.e., of the mapping

$$h: S^{n-1} \to S^{n-1}, \qquad S^{n-1} = \{x \in \mathbf{R}^n : |x| = r\}$$

of a small sphere of radius r centered at O onto itself given by the formula

$$h(x) = \frac{r\mathbf{v}(x)}{|\mathbf{v}(x)|}.$$

Problem 10. Prove that if the operator $\mathbf{v}_{*0}$ of the linear part of the field $\mathbf{v}$ at a singular point O has an inverse, then the index of O equals the degree of $\mathbf{v}_{*0}$.

Problem 11. Find the index of the singular point O of the field in $\mathbf{R}^n$ corresponding to the equation $\dot{x} = -x$.

Ans. $(-1)^n$.

The concept of degree allows us to formulate multidimensional analogues of the two-dimensional theorems considered above. The proofs can be found in books on topology.

In particular, *the sum χ of the indices of the singular points of a vector field defined on a compact manifold of arbitrary dimension is independent of the choice of the field and depends only on the properties of the manifold itself.* The number χ is called the *Euler characteristic* of the manifold. To calculate χ, we need only investigate the singular points of any differential equation defined on the manifold.

Problem 12. Find the Euler characteristic of the sphere S^n, of the projective space $\mathbf{RP}^n$, and of the torus T^n.

Ans. $\chi(S^n) = 2\chi(\mathbf{RP}^n) = 1 + (-1)^n$, $\chi(T^n) = 0$.

Solution. There is a differential equation without singular points on a torus of arbitrary dimension (see e.g., Sec. 24.5), and hence $\chi(T^n) = 0$.

It is clear that $\chi(S^n) = 2\chi(\mathbf{RP}^n)$. In fact, consider the mapping $p: S^n \to \mathbf{R}^{n+1}$ carrying every point of the sphere $S^n \subset \mathbf{R}^{n+1}$ into the line joining the point to the origin of coordinates. The mapping p is locally diffeomorphic, with the preimage of every point of projective space being two diametrically opposite points of the sphere. Therefore every vector field on $\mathbf{RP}^n$ determines a field on S^n with twice as many singular points, where the index of each of the diametrically opposite singular points on the sphere is the same as the index of the corresponding point in projective space.

To calculate $\chi(S^n)$, we define a sphere by the equation $x_0^2 + \cdots + x_n^2 = 1$ in the Euclidean space $\mathbf{R}^{n+1}$ and consider the field $x_0: S^n \to \mathbf{R}$. We then form the differential equation

$$\dot{x} = \operatorname{grad} x_0$$

on the sphere, and investigate its singular points (Fig. 259). The vector field $\operatorname{grad} x_0$ vanishes at two points, the north pole ($x_0 = 1$) and the south pole ($x_0 = -1$). Linearizing the differential equation in neighborhoods of the north and south poles respectively, we get

$$\dot{\xi} = -\xi, \qquad \xi \in \mathbf{R}^n = TS_N^n,$$
$$\dot{\eta} = \eta, \qquad \eta \in \mathbf{R}^n = TS_S^n.$$

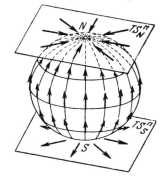

Fig. 259   Linearization of a differential equation on a sphere near its singular points.

Hence the north pole has index $(-1)^n$ and the south pole has index $(+1)^n$, so that $\chi(S^n) = 1 + (-1)^n$.

In particular, it follows that *every vector field on an even-dimensional sphere has at least one singular point*.

Problem 13. Construct a vector field without singular points on the odd-dimensional sphere S^{2n-1}.

Hint. Consider the second-order differential equation $\ddot{x} = -x$, $x \in \mathbf{R}^n$.

Sample Examination Problems

An error of 10–20 % is allowed in all numerical problems.

1. To stop a boat at a dock, a rope is thrown from the boat which is then wound around a post attached to the dock. What is the braking force on the boat if the rope makes 3 turns around the post, if the coefficient of friction of the rope around the post is $\frac{1}{3}$, and if a dockworker pulls at the free end of the rope with a force of 10 kg?

2. Consider the motion

$$\ddot{x} = 1 + 2 \sin x$$

of a pendulum subject to a constant torque. Draw phase curves of the pendulum on the surface of a cylinder. Which motions of the pendulum correspond to the various kinds of curves?

3. Calculate the matrix e^{At}, where A is a given matrix of order 2 or 3.

4. Draw the image of the square $|x_1| \leqslant 1$, $|x_2| \leqslant 1$ and the trajectory of the phase flow of the system

$$\dot{x}_1 = 2x_2, \qquad \dot{x}_2 = x_1 + x_2$$

after time t.

5. Find the number of digits required to write the hundredth term of the sequence 1, 1, 6, 12, 29, 59, ... ($x_n = x_{n-1} + 2x_{n-2} + n$, $x_1 = x_2 = 1$).

6. Draw the phase curve of the system

$$\dot{x} = x - y - z, \qquad \dot{y} = x + y, \qquad \dot{z} = 3x + z$$

going through the point $(1, 0, 0)$.

7. Find all α, β, γ for which the three functions $\sin \alpha t$, $\sin \beta t$, $\sin \gamma t$ are linearly dependent.

8. Draw the trajectory of a point in the plane (x_1, x_2) executing small oscillations

$$\ddot{x}_i = -\frac{\partial U}{\partial x_i}, \qquad U = \frac{1}{2}(5x_1^2 - 8x_1 x_2 + 5x_2^2),$$

subject to the initial conditions

$$x_1 = 1, \qquad x_2 = 0, \qquad \dot{x}_1 = \dot{x}_2 = 0.$$

9. A horizontal force of 100 gm lasting 1 sec acts on an initially stationary mathematical pendulum of length 1 m and weight 1 kg. Find the amplitude (in cm) of the oscillations which result after the force ceases to act.

269

10. Investigate the Lyapunov stability of the null solution of the system

$$\begin{cases} \dot{x}_1 = x_2, \\ \dot{x}_2 = -\omega^2 x_1, \end{cases} \qquad \omega(t) = \begin{cases} 0.4 \text{ for } 2k\pi \leqslant t < (2k+1)\pi, \\ 0.6 \text{ for } (2k-1)\pi \leqslant t < 2k\pi, \end{cases}$$

$k = 0, \pm 1, \pm 2, \ldots$

11. Find all the singular points of the system

$$\dot{x} = xy + 12, \qquad \dot{y} = x^2 + y^2 - 25.$$

Investigate the stability and determine the type of each singular point, and draw the corresponding phase curves.

12. Find all singular points of the system

$$\dot{x} = -\sin y, \qquad \dot{y} = \sin x + \sin y$$

on the torus ($x \bmod 2\pi, y \bmod 2\pi$). Investigate the stability and determine the type of each singular point, and draw the corresponding phase curves.

13. It is known from experience that when light is refracted at the interface between two media, the sines of the angles formed by the incident and refracted rays with the normal to the interface are inversely proportional to the indices of refraction of the media:

$$\frac{\sin \alpha_1}{\sin \alpha_2} = \frac{n_2}{n_1}.$$

Find the form of the light rays in the plane (x, y) if the index of refraction is $n = n(y)$. Study the case $n(y) = 1/y$ (the half-plane $y > 0$ with this index of refraction gives a model of Lobachevskian geometry).

14. Draw the rays emanating in different directions from the origin in a plane with index of refraction $n = n(y) = y^4 - y^2 + 1$.

Comment. The solution of this problem explains the phenomenon of the mirage. The index of refraction of air over a desert has a maximum at a certain height, since the air is more rarefied at higher and lower (hot) layers and the index of refraction is inversely proportional to the velocity of light. The oscillations of the ray near the layer with maximum index of refraction is interpreted as a mirage.

Another phenomenon explained by the same kind of ray oscillations is that of acoustic channels in the ocean, along which sound can be propagated for hundreds of kilometers. The reason for this phenomenon is the interplay of temperature and pressure leading to the formation of a layer of maximum index of refraction (i.e., minimum sound velocity) at a depth of 500–1000 m.

An acoustic channel can be used, for example, to give warning of tidal waves.

15. Draw geodesics on a torus, using Clairaut's theorem which states that the product of the distance from the axis of revolution and the sine of the angle made by the geodesic with a meridian is constant along every geodesic on a surface of revolution.

Bibliography

Bellman, R., *Stability Theory of Differential Equations*, McGraw-Hill, New York (1953).

Birkhoff, G., and G.-C. Rota, *Ordinary Differential Equations*, Ginn, Boston (1962).

Coddington, E. A., and N. Levinson, *Theory of Ordinary Differential Equations*, McGraw-Hill, New York (1955).

Hartman, P., *Ordinary Differential Equations*, Wiley, New York (1964).

Hurewicz, W., *Lectures on Ordinary Differential Equations*, MIT Press, Cambridge, Mass. (1958).

Ince, E. L., *Ordinary Differential Equations*, Dover, New York (1956).

Lefschetz, S., *Differential Equations: Geometric Theory*, second edition, Wiley (Interscience), New York (1962).

Nemytski, V. V., and V. V. Stepanov, *Qualitative Theory of Differential Equations*, Princeton University Press, Princeton, N.J. (1960).

Petrovski, I. G., *Ordinary Differential Equations* (translated by R. A. Silverman), Dover, New York (1973).

Pontryagin, L. S., *Ordinary Differential Equations* (translated by L. Kacinskas and W. B. Counts), Addison-Wesley, Reading, Mass. (1962).

Simmons, G. F., *Differential Equations with Applications and Historical Notes*, McGraw-Hill, New York (1972).

Index